…e that must be prioritized for the developing world. With …ng understanding of these issues, and her deep knowledge …al and local business opportunities, Naina has put together … valuable compilation of how we can be more prudent and …t about utilizing our natural resources to realize a future that …lthier and safer.'

—MELINDA GATES,
Co-Chair, Bill & Melinda Gates Foundation

…order to accelerate the transition to a low-carbon economy, we must …d the right policies for each particular context. Luckily, there are now … wider range of policy options—some of which could potentially be …ontroversial—that we can choose from to achieve our environmental …oals while reaping short-term benefits for our communities. This book precisely contributes to show that there are coalitions which can be built and complex problems which should be addressed by strong leadership. Moreover, Naina points to water and sanitation as two action areas where there is huge potential as a means to fully transition to a low-carbon economy.'

—FELIPE CALDERÓN HINOJOSA,
Former President of Mexico; Honorary Chair of the
Global Commission on the Economy and Climate

'Naina brings long and broad experience, including prominently in finance, and a holistic approach to the huge challenges and opportunities of sustainable development in India. Providing the public and private financial resources at the necessary scale and pace requires better incentives and rigorous risk disclosure, and innovative financial instruments and institutions that India's financial sector and policymakers are well positioned to develop in concert with partners around the world. Naina plays a leading role globally in our joint journey of learning and advocacy for sustainable development and a new climate economy.'

—CAIO KOCH-WESER,
Chairman, European Climate Foundation;
Former Vice Chairman, Deutsche Bank;
Former German Deputy Minister of Finance

'Naina has brought her undoubted leadership, passion and clear thinking that she demonstrated in her highly successful career in financial services to the challenging and complex world of sustainable development. Her well-articulated ideas and thought pieces on the topics of water and sanitation—supported with on the ground examples of different types of organizations playing a critical role in addressing these twin challenges—is a timely "must-read" for policymakers, corporations and individuals who want to make a difference.'

—**RICH LESSER,**
President and CEO, The Boston Consulting Group

'It is difficult to imagine India being a developed country without solving its problems in water and sanitation. Naina has seen the issue both as a corporate leader and a responsible citizen. In her book, she has explored the interlinkages of water and sanitation with social development in great depth. The issues and solutions covered in the narrative make the book a very good primer for anyone who wants to drive positive change.'

—**ANAND MAHINDRA,**
Chairman and Managing Director, Mahindra Group

'We all understand the need for better sanitation and water management and a cleaner environment which provide green jobs and inclusive growth. This book combines successful initiatives and personal experiences to weave them into institutional frameworks that allow us to see how these goals can be achieved.'

—**AJAY MATHUR,**
Director General, The Energy and Resources Institute (TERI)

'Naina Kidwai has a unique view to the world's wicked problem of lack of access to sanitation, which affects the dignity and health of the poorest. She comes to the issue as a banker and a doer, and so provides fresh insights into what will work and how. This is important as the world is struggling to find answers to sustainable growth, which has to be inclusive and affordable to be sustainable.'

—**SUNITA NARAIN,**
Director General, Centre for Science and Environment

'Beyond her brilliant corporate career, Naina has always taken keen interest in matters of water, sanitation and the environment. Her understanding of both business and environmental issues has led her to give meaningful ideas on a path to a green economy with job creation. This book is a great read for anyone trying to understand these complex issues in a holistic manner and looking for pragmatic outcomes.'

—NANDAN NILEKANI,
Co-founder and Non-executive Chairman of Infosys;
Founding Chairman of Unique Identification Authority of
India (UIDAI); Co-founder and Chairman of EkStep

'Nowhere is the simultaneous challenge and opportunity of sustainable development more evident than in India. In this book, Naina shows the interdependency of human development with the fight against climate change and environmental destruction. Drawing on her many years of experience in finance and her personal experiences in India, she shows how we can harness the opportunities for new jobs and a new kind of growth that will accelerate our path to a brighter future for all.'

—PAUL POLMAN,
CEO, Unilever

'From defining the role of corporates and communities in sanitation to a multifaceted approach to solving India's water woes, Naina Lal Kidwai's path-breaking book shows us the way to a strong and ecologically sustainable future. As the head of the India Sanitation Coalition, I cannot think of anybody more capable than Naina herself to educate us all on issues as relevant as climate economy, green jobs and the significant role of women in social transformation. We are all certain to benefit immensely from extraordinary effort.'

—SURESH PRABHU,
Minister of Commerce and Industry, Government of India

'Water is emblematic of the most urgent and important problems that humanity must tackle. It is a maze of technical, economic, political, social and ethical challenges. There is a compelling need for leaders to engage with these problems—intellectually and in action—Naina does this comprehensively through this book and sets an example for others.'

—AZIM PREMJI,
Chairman, Wipro

'Naina Kidwai brings her vast experience from years of working in finance and public service and from participating in domestic and international debates to explain the problem of water use in India: why it is so important, how we are falling behind, and what we need to do. There is no water shortage in India, and little difficulty in keeping our rivers flowing and clean, in getting rid of the scourge of diarrhoea and other waterborne ailments, in disposing of waste profitably, and in sparing our people the indignity of open defecation if we follow through on some of the important suggestions in this book. Naina's voice is informed and passionate and well worth hearing.'

—RAGHURAM G. RAJAN,
Professor of Finance, University of
Chicago Booth School of Business

'In this pioneering book, Naina Lal Kidwai draws from a lifetime of experience to cast a sharp eye on important dimensions of India's environmental challenges. Practical examples highlight the roles of different partners, including the private sector, faith-based organizations and, in particular, the critical role that women play in helping to address them. Her insights are sure to stimulate new thinking on critical issues for sustainable development in India and around the world.'

—ACHIM STEINER,
Administrator, United Nations Development Programme

'Dirty water and poor sanitation take a terrible toll in India, which is home to 30 per cent of the world's deaths from diarrhoea. But the quality of water and sanitation can change rapidly and, while there is a long way to go, such change is coming, in large measure as a result of the work over many years by people such as Naina Lal Kidwai.

These improvements require both investment and behaviour change and she writes very thoughtfully about both. Women suffer the most from these problems, but they are also key agents of change. Further, she argues clearly, with her great private sector and NGO experience, the importance of collaboration among the government, private sector and NGOs.

She also writes strongly and persuasively about climate change and growth, showing that the only way to grow sustainably is also to be responsible about the climate. Her writing, too, on wildlife is very powerful, again embodying a lifetime commitment. This is not only a very valuable piece of work, it also represents decades of tireless effort to change the world.'

**—LORD NICHOLAS STERN OF BRENTFORD,
CH, FBA, FRS; IG Patel Professor of
Economics and Government and Chair of the Grantham
Research Institute on Climate Change and the Environment
at the London School of Economics;
Former Chief Economist at both the World Bank and
European Bank for Reconstruction and Development**

SURVIVE OR SINK

SURVIVE OR SINK

An Action Agenda for Sanitation, Water, Pollution and Green Finance

NAINA LAL KIDWAI

Published by
Rupa Publications India Pvt. Ltd 2018
7/16, Ansari Road, Daryaganj
New Delhi 110002

Sales centres:
Allahabad Bengaluru Chennai
Hyderabad Jaipur Kathmandu
Kolkata Mumbai

Copyright © Naina Lal Kidwai 2018

The views and opinions expressed in this book are the author's own and the facts are as reported by her which have been verified to the extent possible, and the publishers are not in any way liable for the same.

All rights reserved.
No part of this publication may be reproduced, transmitted, or stored in a retrieval system, in any form or by any means, electronic, mechanical, photocopying, recording or otherwise, without the prior permission of the publisher.

ISBN: 978-81-291-5140-7

First impression 2018

10 9 8 7 6 5 4 3 2 1

The moral right of the author has been asserted.

Printed in India by Replika Press Pvt. Ltd.

This book is sold subject to the condition that it shall not, by way of trade or otherwise, be lent, resold, hired out, or otherwise circulated, without the publisher's prior consent, in any form of binding or cover other than that in which it is published.

To Rumaan, Kemaya and Zahaan
In the hope that we will leave you a cleaner, better world that you will sustain and conserve for future generations

Contents

Foreword by Amitabh Kant   *xv*
Introduction   *xxi*

Water and Sanitation

Sustainable Sanitation: A Review	3
Galvanizing the Power of Multiples	7
Corporates as Partners in Achieving a Swachh Bharat	13
Swachhata Doots: The Power of Volunteering	39
Role of NGOs and Development Agencies in Sanitation	44
German Expertise in India's Rise as a Swachh Nation	56
Faith and Sanitation	69
Move to Sanitation Plus	74
Effective Waste Management Is the Need of the Hour	78
A Case for Faecal Sludge Treatment Plants	82
Making Every Drop Count	92
Role of Corporates in Water Stewardship	99

Environment and Pollution

Collective Action on Climate Change	109
In the New Climate Economy	113
The World We Are Yet to Build	119
Government Is Key, Business the Solution	131

Rivals and Partners: China and India	136
Addressing Pollution	141
Wildlife and Habitat Conservation	148

Green Economy and Finance

The Future Path for Green Finance	161
Financing Green Energy in a Power-starved Country	167
Financing India's Urban Development Pathway	173
Global Finance and Global Warming	185
Building the BRICS of the Paris Agreement	188
The Case for a BRICS Rating Agency	191
Chasing Green Jobs	196

Women

Women's Empowerment: Nurturing the Ecosystem	207
Rural Indian Women: Sowing the Seeds of Social Transformation	219
Women in the Sustainability Discourse: Time for Action	230

Epilogue	237
Acknowledgements	243
Glossary of Important Terms	245
About the Author	249
Index	251

Foreword

Amitabh Kant, CEO, NITI Aayog*

For a young nation with immense potential and a vision of development that will vanquish poverty as well as fulfil the highest aspirations of its diverse population, the importance of accelerated but sustainable development must go beyond rhetoric, to concrete action. Naina throws the spotlight on this and sets the agenda for several core themes of sustainable development, including water, sanitation, environment, financing, green jobs and the role of women in social transformation. This collection of articles reflects a comprehensive understanding of issues on the ground, while being able to link them to effective strategies requiring the active engagement of citizens, businesses and the government.

While adopting the Sustainable Development Goals (SDGs), the global community acknowledged that in order to accelerate development, a more comprehensive approach that links economic, social and environmental issues is critical. At NITI Aayog, there is a distinct focus on catalyzing transformative action in the social sector. Appreciating that in

---

*The National Institution for Transforming India (NITI Aayog) is the premier policy think tank of the Government of India, providing both directional and policy inputs.

a federal structure implementation is primarily the prerogative of state governments, cooperative federalism has been taken up a notch to an innovative form of competitive federalism. Through this approach states are nudged towards improving their development outcomes, with NITI measuring annual changes in performance. States are also ranked on the basis of these changes. Indices have been developed for measuring outcomes in health, education as well as water management and sanitation. Each index contains a composite set of indicators, with maximum weightage accorded to outcome indicators.

The Water Index, for instance, aims to measure the sustainable management of water resources. It assesses states on indicators pertaining to source augmentation, watershed development, participatory irrigation practices, drinking water supply and sanitation, among others. The Health Index measures key and intermediate outcomes in maternal and child health, infectious diseases, out-of-pocket expenditure, governance, data integrity and selected health system inputs, including human resources and financing. Health being an outcome of multiple determinants, the Index aims to elicit action across government departments, so that innovative interventions and corrective steps are implemented. Another focus area for the social sector development agenda at NITI is school education and the NITI School Education Quality Index measures learning outcomes, access, equity and governance processes.

These indices have been developed through an iterative consultative process with the concerned central ministries and state governments, with a view to bringing a renewed focus on accelerating progress in the social sector, which has historically been accorded lower priority. It also ensures that states are ranked on the basis of change in performance from one year to the next, rather than historical achievements, thereby encouraging those traditionally at the bottom of the 'merit list' to step up efforts. The Ministry of Health has linked performance on the Health Index with a financial incentive to states under the National Health Mission, further leveraging its potential for propelling transformation.

Further, the Swachh Bharat Mission (SBM) remains a priority area for the government. The National Family Health Survey 2015–16 reports that while 90 per cent of households have access to an improved drinking water source, only 48 per cent use an improved sanitation facility. We still have a long way to go. Naina's work with the India Sanitation Coalition (ISC) has been significant in bringing together stakeholders as well as several independent initiatives on to a common platform. This helps to aid cross learning and thereby facilitates more effective actions for sustainable interventions in the WASH (Water, Sanitation and Hygiene) space.

The book highlights learnings from these initiatives, including the importance of moving beyond toilet construction to effecting individual behaviour change in rural areas, while urban areas pose contrasting challenges. The need for altering behavioural practices is a particularly important issue because it makes the difference between successful and unsuccessful initiatives in water and sanitation. Instances of people drinking water from contaminated sources have been recorded even when potable water is supplied by a government agency. Further, Naina brings out the important role played by faith-based organizations (FBOs) and local community leaders. In order to be effective, behaviour change campaigns need to be tailored to local communities. Evidence suggests that for many households, health is not as strong a motivating factor for toilet construction as compared to social pressure.

Infrastructure and technology required for the treatment of waste and mechanisms for converting waste to wealth are also highlighted. The section on water management and stewardship presents ideas around pricing and creating a market for treated municipal water. Case studies of companies that are achieving or have the potential to achieve a positive water balance have also been included. The book contains important examples of how collaboration among corporates, non-governmental organizations (NGOs), citizens and the government can bring about transformative action in the area of water and sanitation. Importantly, the book highlights the impact of poor water and sanitation on causing a wide range of diseases, propagating the cycle of malnutrition and ultimately reducing the productivity of the nation.

In 2015, the Sub-Group of Chief Ministers on Swachh Bharat Abhiyan constituted by NITI Aayog put forward several recommendations, the implementation status of which is being monitored across ministries on a regular basis. A number of these measures have already been implemented. These include the Swachh Bharat Cess, compulsory procurement of 100 per cent power produced from all waste-to-energy plants and market development assistance of ₹1,500 per tonne of city compost. A special one-time assistance of ₹1,000 crore, recommended by NITI, has been released. These funds will be utilized for setting up community water purification plants in nineteen arsenic and fluoride affected states as well as for surface water projects in Rajasthan and West Bengal.

The book in its next section focuses on environment and India's role in the world, wildlife and forests. By committing to reduce the country's emissions intensity per unit of Gross Domestic Product (GDP) under the National Determined

Contributions, we have demonstrated our responsibility and leadership in this area. India also plans to create an additional carbon sink of 2.5–3 billion tonnes through additional tree cover. The book describes solutions for building a sustainable future through an emphasis on cities. Examples of businesses embracing the climate change agenda by taking responsible actions have also been highlighted.

Two vital enablers for the sustainable development agenda—financing and green jobs—are also discussed in the book. Among other initiatives, the need for Indian banks to adopt standards for financing projects on the basis of their environmental responsibility has been deliberated upon. Green jobs that have the potential to transform must be made aspirational, as Naina explains. She provides examples of entrepreneurial initiatives in the waste management sector that bring together vocational skills as well as the ability to innovatively organize, manage and implement models that bring essential services to people.

The final theme in the book is the role of women in social transformation. This cuts across a number of the aforementioned themes by providing examples of how women have led sustainable development initiatives in their homes and communities. Recognizing that several such initiatives often go unnoticed, NITI launched the Women Transforming India Awards in 2016. This initiative provides a platform for sharing the success stories of remarkable women across the country. Work is also under way to develop an index that can measure various elements of gender equality, including economic participation, political voice, health and education, security and justice as well as access to household infrastructure, among others.

In conclusion, there is no denying that government leadership is critical for bolstering the agenda for human development and environmental sustainability. NITI Aayog, with the Prime Minister as its Chairperson, is responsible for providing overall coordination and leadership in achieving the 2030 Sustainable Development Agenda. A delegation led by the Vice Chairman, NITI Aayog, recently presented India's Voluntary National Review encapsulating the progress made by the country on various SDGs at the United Nations High-Level Political Forum. In addition to government leadership, innovation and creative thinking that involves the participation of citizens is indispensable. Collective problem-solving and developing innovative measures can provide the much-needed push for accelerating India's development momentum. While several measures have already been taken, many others remain. In this book, through her personal contributions to the social development space while pursuing her corporate career, Naina brings this very point to fruition, and inspires the reader to think of his or her own role in India's shared development story.

Introduction

I have enjoyed birdwatching on the banks of the Yamuna in Delhi in my college days—migratory birds would stop over and waders would dabble by its shores. Today, forty years later, this same river is a sewer for Delhi's untreated waste. As an erstwhile banker, I have celebrated and revelled in India's progress and growth. But this cannot be at the cost of destroying what we have—our air, water, forests and wildlife, which are our lifelines. The good news is that India is blessed with a criss-cross of rivers and waterbodies, arguably with sufficient water if respected and used judiciously. That we have destroyed so much of our natural wealth is the tragedy. The poor have continued to suffer as we have not been able to deliver water so they pay more than they should at the hands of greedy entrenched water mafias.

Over the last ten years, I have been deeply engaged in issues concerning water and environment both in India and abroad. With a view to improving corporate engagement in water efficiency and water stewardship, I started the Water Mission at Federation of Indian Chambers of Commerce and Industry (FICCI), instituting awards and producing compendiums of best practice. I learnt from these studies and many projects on the ground in India. While at the Hongkong and Shanghai Banking Corporation (HSBC), we were deeply engaged in environment and water projects working through NGOs operating in India. As a board member of Shakti Sustainable Energy Foundation and The Energy and Resources Institute (TERI), I was exposed to frameworks and policy requirements and issues in these areas in India. As Co-Chair of the Advisory Council on Water at the World Economic Forum and as a Global Commissioner of the New Climate Economy (NCE), I was exposed to global thinking on these topics. My role as a member of the International Advisory Council for the United Nations Environment Program (UNEP) Inquiry for Sustainable Finance and Participation at Davos on these subjects brought me into the conversations on financing. As Chair of the UNEP Inquiry India Council, we developed some of the thinking around financing India's needs. At various stages, I wrote articles and op-eds for various publications.

On suggestions from various people, I have attempted to compile some of these articles of the last couple of years into this book. The danger of something like this is that the work is not comprehensive and some issues get left out—these I will attempt to highlight in this introduction.

SANITATION: THE SBM AGENDA

A brainstorming on sanitation with some experts led my husband and me to start the India Sanitation Coalition to bring all players in the sector together to collaborate—providing a platform for NGOs, development partners, corporates, donors and the government. We had learnt of four NGOs working in the same district in Bihar, but none of them knew what the other was doing and there was little sharing of common learnings with each re-inventing the wheel. The need for collaboration was critical as all actions in sanitation were sub-scale—more so given the huge agenda set out by Prime Minister Narendra

Modi. The SBM was launched on 2 October 2014 by Prime Minister Modi with a view to a clean India, just after his first speech on India's Independence Day. No one doubts the conviction and personal leadership with which he drives this agenda even today.

WHY IS SANITATION SO CRITICAL?

When a jumbo jet crashes, it makes front-page news for days. Diarrhoea accounts for 1,600 deaths daily—the same as eight jumbo jets crashing every day. But this is not considered newsworthy and barely gets our attention. Here are some more figures that we must consider:

- Thirty per cent of childhood diarrhoea deaths are in India
- Three children under the age of 5 die every minute in India
- Unsafe water is a major cause of this, due to chemicals (fluoride, arsenic) and bacteria
- Twenty-five per cent of the country is still defecating in the open with no toilet or sewage system connections

India generates 1.75 million tonnes of faecal waste every day. However, there are no systems in place to safely dispose of this waste. Nearly 75 per cent of this sludge—human excreta and water mixture that bears disease-carrying bacteria and pathogens—remains untreated and is dumped into drains, lakes and rivers posing a serious threat to health. This contaminates our water table as it seeps untreated into our waterbodies and contaminates the food we eat as flies and insects carry disease when they flit between the faeces and our food.

India has the highest incidence of stunting where 40 per cent of children do not grow to their full physical level and even worse, their brains remain underdeveloped—so much for the demographic dividend! The good news is that by October 2016, India had moved from 40 per cent to 67 per cent safe sanitation. We still have a long way to go, but there is no doubt that the momentum has picked up. In citizen surveys, the SBM programme gets voted again and again as the most noteworthy programme of the Modi government. Seven states—Sikkim, Kerala, Punjab, Himachal Pradesh, Uttarakhand, Haryana, Arunachal Pradesh—and NDMC Delhi (area under the New Delhi Municipal Corporation) were declared Open Defecation Free (ODF) in 2017, and many others will follow.

The social benefit of sanitation as per the World Bank is $53 billion. The SBM is critical for India to reap the demographic dividend that economists talk about—a young and healthy India benefitting from the low dependency ratios many countries would love to have.

SANITATION AND BEHAVIOUR CHANGE

Though toilets are a must for sanitation, the government, fortunately, has course corrected and we have moved from building toilets to now focusing on behaviour change. Till the ills of open defecation are accepted, a very large number of our people will continue to defecate in the open as they have done for centuries. Many toilets were built particularly in rural India with the help of subsidies provided by the government, and then were used as storerooms. We need people to understand why using the toilet is important. Women were early acceptors for reasons of security as they moved at dawn and dusk to fields and were sometimes attacked, even raped, en route, as darkness spread around. Urinary infections caused by holding back have a high incidence in such areas where toilets are not available. But to be ODF and prevent the hazards of defecating in the open, men have to embrace the cause as well, as the health of the entire village depends on them too. Toilets that are built need to be used and maintained. Schools have embraced the building of toilets with great alacrity, recognizing also the essential requirement of separate toilets for girls to ensure they stayed in school. Hence, the commitment of the ISC is to Build, Use, Maintain and Treat (BUMT).

Also outlined in the book are corporate volunteering programmes like Swachhata Doots, where factory employees (with some training and a mobile app) are effecting behaviour change in communities around the factory.

THE ROLE OF FAITH

The role of faith through FBOs in this course of behaviour change is critical. Faith is the magic wand to create a Swachchata Kranti—a cleanliness revolution as people accept, believe and follow their faith leaders whose outreach and influence is enormous. We should celebrate our Swachhata Karmis like we celebrate our soldiers. It is time for a change in mindsets.

Islam says we must work for humanity. It talks of Amana—custodianship and accountability. We must protect whatever resources (water, forests) we have. 'Do not waste water, even if you are at a running stream,' said Prophet Mohammed.

Pope Francis revisited the canticle of the creatures and redefined and reminded us about the importance of the environment and water. Five global commissioners (including myself) from the NCE, spent a day with Pope Francis and archbishops from around the world on the subject of environment and conservation a couple of months before the publication of his ecology encyclical—*Laudato Si*. It turns out the Vatican had been working on their message on conservation and respecting nature for a couple of years before the declaration.

The outreach of our swamis, gurus, pirs, sants and priests has an enormous impact on how we respect nature and the environment and as we look to change the way we live and protect our resources. Fortunately, there are ways to ensure we do not pollute water and use it efficiently and with care.

ROLE OF NGOs AND DEVELOPMENT AGENCIES IN SANITATION

In this chapter, I have given examples of how these players are critical in delivering results for an ambitious programme such as SBM. Sustained collaboration between government and NGOs enables the programme to leverage respective strengths and resources and ensures the longevity of results.

WASTE TREATMENT

If we do not tackle treatment of faecal sludge, we would have to embark on one of the most expensive exercises in history to remove it from our fields by funnelling it into toilets and then only putting it right back in our fields. The chapter on faecal sludge management (FSM) looks at the ways we must address the treatment of waste—the technology and the infrastructure. Yet, here too, we need behaviour change that elevates the skills and jobs that deal with this, and that removes the bias against buying and using fertilizer and waste water that emanates post treatment. The twin pit technology may well emerge as the gold standard for simple, low-cost and effective treatment. The chapter on faecal sludge treatment plants (FSTPs) looks at one possible solution for urban India where distributed systems may provide the answer to deal with untreated waste.

We need to recognize waste is wealth. The market for FSM compost is resistant, but this offtake is important. Farmers must pay for treated faecal sludge to be dumped in their farms (as in some states already) and need to recognize that this fertilizer helps non-food crops like cotton.

Similarly, waste water after treatment often gets released into our rivers, rather than being used for horticulture or sold to power plants to use for cooling purposes. We need a price to be established—a market for treated waste water and treated faecal waste.

The role of countries such as Germany in sanitation is important for us to learn from and indeed German companies and their local governments are engaging with technology and money. Germany is one of the largest donors in the sector. The GIZ[1] supported Sustainable Sanitation Alliance (SuSanA) has an Indian chapter now with the ISC increasing and encouraging the conversations and understanding of sanitation and linking us to global experts.

Large companies embrace these ideas and comply with regulation. Managing these issues with small and medium sized enterprises (SMEs) is much more of a challenge, as increased capital costs need funding and the fear that passing these costs on to consumers because of the potentially increased price of end products and rendering them uncompetitive, keeps them away from doing what is required. If we can get the leather industry, sugar plants and mini paper mills to comply, we will solve industrial pollution of our rivers in the Gangetic river basin.

URBAN SANITATION

The chapter on corporates and urban sanitation touches on the challenge of increasing Corporate Social Responsibility (CSR) spends for urban solutions as corporates have preferred rural

---

[1]The Deutsche Gesellschaft für Internationale Zusammenarbeit (GIZ, formally known as GTZ) is a federally owned organization. It works worldwide in the field of international cooperation for sustainable development. Their mandate is to support the German Government in achieving its development objectives.

interventions. The Ministry of Urban Development (MoUD) recently launched the Swachh portal to enable a working interface for collaboration between Urban Local Bodies (ULBs) and corporates along with entrepreneurs in the sanitation value chain. The issues for behaviour change are quite different as people are clamouring for toilets in urban environments. We need to solve the problem in urban slums where 15–20 per cent of our city's population resides and where provision of toilets is a challenge because of space constraints and non-availability of sewage pipes. The challenges of operating community toilets are formidable. Fortunately, there are some successful models, e.g. the Society for the Promotion of Area Resource Centers (SPARC) model in Dharavi slums (Maharashtra), Sulabh International's pay-to-use models and the recent toilet and laundry Suvidha facility put up by Hindustan Unilever Limited (HUL) in Ghatkopar (Maharashtra). We need more such initiatives.

RURAL SANITATION
Sanitation in rural India has been easier to tackle with panchayats and district administrations taking the lead, often supported by NGOs and corporates. The dynamic secretary of the Ministry of Drinking Water and Sanitation (MDWS), Parameswaran Iyer, has correctly recognized the twin pit toilet as the gold standard, as it is cost-effective and not dependent on water flushing mechanisms. However, innovations are continuously required to improve design, adapting these to different locations. The supply chain also needs to be strengthened to reach all locations.

The issue of safe and sustainable sanitation is both a health and development imperative. It is critical to align with the country's goals and achieve the resulting outcomes of better health, education (as attendance of children, especially girls, improves) and community welfare. The current focus on sanitation, which is historically unprecedented, provides an opportunity to create strategies and business models for delivering effective, affordable and sustainable WASH solutions suitable for the Indian context.

WATER

Sanitation and water are firmly interlinked. In fact, my interest in water took me down the stream to sanitation. You cannot have clean rivers without treating sewage. You cannot have pure waterbodies without dealing with sanitation. And you cannot have sanitation without water, as technologies today need water to clean toilets. We certainly need technologies that minimize water disrupters like biodigesters. But till these prove they are economic and can be scaled up, water remains essential in the sanitation space.

The United Nations (UN) has ranked India as 120th of 122 countries for water quality estimating that 70 per cent of our water is contaminated. To compound this, as reported by World Resources Institute (WRI), 54 per cent of India's total area is under high to extremely high water stress and groundwater is declining in 54 per cent of wells across India. The chapter, 'Making Every Drop Count', highlights the scale of the problem.

Resource mismanagement, underdeveloped infrastructure, poor technology and unequal governance structures are at the heart of India's water and food security. Efforts on water sustainability need solutions that address food and agricultural practices, climate, industry and ecosystems that depend on water. We need to propose realistic pricing of water, as this will encourage investment in water treatment and water conservation methods. Pricing and creation of a market for treated municipal water needs to be incentivized. We need policies and punishment mechanisms around water management. Some suggestions are mentioned in the chapter.

The chapter on water stewardship examines shared risks and offers an opportunity to harness the shared value of water. This is a new concept gaining acceptance in India and highlights how some companies have managed to reduce their water footprint, even achieving a positive water balance by replenishing more than they extract.

Water stewardship starts with companies responding to their own water related risks through improvement in policies and processes, collaboration with external stakeholders, including its supply chain, to understand and minimize potential impacts and work collectively for enhancement of efficient water management including at the catchment level. There are examples of water stewardship initiatives of several companies, including measurement and voluntary disclosure initiatives on water footprint. I hope over time such reports become mandatory and ideally prepared and vetted by specialists.

To encourage water stewardship, there is a role for government in promoting these policies; financial institutions in ensuring compliance and measuring water related risks; civil society and government in supporting commitment to accountability, developing and implementing policies and educating consumers; and consumers by their purchasing decisions and policy advocacy.

COLLABORATION ACROSS STAKEHOLDERS

A number of chapters highlight success stories where communities, government, development organizations and civil society have come together to provide safe water to their citizens. Thus, we succeed when all players come together and where true collaboration and partnerships exist.

Both for sanitation and water, citizen action is essential and communication through government and NGOs is key. The programmes which endure are where communities are organized around the problem and own the solution. District collectors/government representatives come and go, but the community is deeply anchored at the location.

The role of corporates as a stakeholder is also important. Several articles on sanitation focus on the way CSR monies are directed into sanitation and explore the companies' actions. The role of corporates in sanitation and water brings good execution of projects on the ground, innovative approaches and funding. There are great examples of success where collaboration across stakeholders produced the desired result. However, to make this happen, you need strong leadership. In the example of Moradabad, an NGO (World Wildlife Fund, or WWF), an effective district collector and a corporate (HSBC) worked with the community to create awareness of how metal poisoning of the water was affecting their health. The project was closely monitored and outcomes measured. Outreach into hundreds of small brassware and metal craft companies helped bring the SMEs on board to ensure a major clean-up of the water, and also control the effluents flowing into the river.

In all this, the government, particularly at the state level, is a key player given the scale of operation required. Again, the role of partnerships and collaboration is important. Excellent examples include the Maharashtra government's 1,000 village scheme where a Section 8 company has been established with corporates contributing funding and expertise and the government matching the funds. Implementation is being overseen by the Chief Minister's office.

Another example is the Swachh Bharat Fellows where Tata

Trusts have funded ₹600 million and are recruiting and training individuals who will be present at district collector offices to support sanitation initiatives. This programme was developed with the MDWS, and is one of the biggest such collaborations between the government and the private sector.

There are very few examples of private sector ownership of city systems, although I believe Bangkok Sewerage and Manila Water are good examples of private ownership in delivery of public goods.

POLLUTION

Pollution is a raging issue for cities in India. Before we despair at the ever deteriorating air quality in Delhi and its environs, we must remember that such situations have prevailed in other major cities like London and Los Angeles and have been tackled successfully. Beijing too has made progress. This did not happen overnight but took several years. It required working to a well-thought-out comprehensive plan. We can, and must, do so too.

The chapter on pollution looks at the several causes and possible solutions touching on the way we generate power, transportation and emission standards, industrial pollution, brick kilns, dust, waste treatment and agricultural practices. The chapter also argues for the need to measure and track pollution against standards, so as not to lose momentum.

Resolving the problem takes time, often many years. Remedial measures initiated require follow through on a consistent and continuous basis. There needs to be a multi-year, year-round plan of action, and not a series of reactions each time air quality deteriorates on a seasonal basis. The fundamental elements of poor air are with us throughout the year, with some seasonal variations and spikes. We need to address these root causes on a long-term basis.

FUNDING

Finance is a critical enabler as without it little can move forward. In the chapters on green economy and finance, I argue that we need a sustainable finance architecture where everything we do is seen through the lens of sustainability. Banks should not fund projects which are not doing the needful to contain effluents or that pollute. A banking code needs to be operational and adhered to across banks to fund what is right and not fund what is not. Fortunately, the first steps in this direction are evolving, although India remains a laggard as our banks have yet to develop a common understanding and measurement in this regard. The global investor community already applies these principles when they look to invest in companies, and Indian firms that access global capital recognize this. We need a benchmark for domestic capital too as finance can discipline companies to do what is right and, at a minimum, adhere to local regulations. Credit decisions must factor in risks—30 per cent of our power plants shut down in summer for lack of water. The West is replete with cases of old coal-based plants having shut down as communities see the environmental risk and health implications from pollution and these 'stranded assets' become a problem for lenders.

We need vibrant capital markets and regulatory practices which enable our municipalities to raise funds. We need credit-enhancing mechanisms which de-risk projects, enabling funds to be raised and lowering the cost of funding. In the chapters in this section, the need for finance to achieve our ambitious renewable energy agenda is discussed.

We need to mix and match public and private resources, producing a blended rate of finance which is lower than market rates to make public services more affordable. Government financing needs to be speedy, dependable and transparent, and should focus subsidies to the poorest. De-risking by public monies will allow entry of private finance. We need new capital market instruments like green bonds and Invescos.

The role of microfinance and last-mile finance to close the gap even where subsidies are available is a proven way to aid citizens in their aspiration to build their own toilets. The Reserve Bank of India (RBI) has included small loans financing for sanitation in priority sector lending by banks, but little has moved in this area as banks are looking for aggregators as they cannot administer multiple micro loans. The southern states have shown the way and the ISC and Water.org are working to introduce microfinance and bank funding for sanitation into other states.

SKILLING AND GREEN JOBS
The role of government and the National Skill Development Corporation (NSDC) is paramount in the skilling required to deliver sanitation, water and renewables infrastructure. The chapter, 'Chasing Green Jobs', highlights that solar and wind energy have created 70,000 jobs thus far. If India achieves its target of 100 GW of solar energy by 2022, as much as one million jobs would be created. Studies conducted by the Skill Council of Green Jobs (SCGJ) indicate that sixty-five million jobs will need to be created by 2030, with the largest potential being in waste management and water management, together accounting for 30 per cent. We also need to make the work aspirational and value add to their skills. The skill councils are working on the occupational standards while looking at the differences for each state. We need masons, plumbers and engineers and training for manpower at FSTPs to manage and

maintain plants and the desludging required after three years. Ragpicker skills require to be upgraded even as we need to create a mindset change amongst workers.

A key approach to skilling could come from creating entrepreneurs, required where the training is not just about a vocational skill, but also on a business model in which a service or product is delivered for a profit. This will ensure that a million entrepreneurs bloom. In order to deal with waste, we will have to show the way by demonstrating successful waste-to-wealth models.

THE ROLE OF ENTREPRENEURSHIP
The NSDC has promoted the Nirvana Fund with business leader Subroto Bagchi for Odisha, i.e. the fund to act as angel investors. We need venture and start-up capital to encourage new ideas. This will ensure engagement with the customer even in the absence of government support. There are drinking water models in place today where small plants treat and then charge for water provided by them to the community, providing a vital public service albeit at a cost.

In the waste treatment mandated by the law, we have compliance by large buildings and institutions. In Kampala, Uganda, ten millionaires were created from a business model where they collect waste from slums for a small cost in colour-coded plastic drums that slide under the toilet. These are then transported to a plant that treats the waste to produce fertilizer for cotton crops in surrounding fields. Such simple and effective ideas need technology, a local municipality which encourages the existence of the entrepreneur and, as in this case, an NGO which helps connect the dots. Likewise, there are women who run the cleaning of community toilets in Dharavi, revenue of which is ₹10 million a year through servicing models worked out with the NGO, SPARC. Small but significant interventions like charging a small amount from each family, giving the toilet bank cleaner his home in the vicinity of the toilets, thereby ensuring cleanliness levels; a women's group of office bearers created from the community to monitor and manage; and a municipality providing the space and the sewer connection for the toilet bank—all make for a successful model. We need to replicate what works by educating and training people to implement these models.

THE ENVIRONMENT AND INDIA'S ROLE IN THE WORLD

In the chapters on environment and policy, I look at India's chance to lead. It is estimated that pollution costs India the equivalent of 5.5–5.7 per cent of the GDP each year. Half of the most polluted cities of the world are in India. Our energy costs are doubling every fifteen years—we spent an average of 6.5 per cent of our GDP on importing fuel from overseas. Excessive drawdown of groundwater for agriculture and the costs of poor urban planning are very costly growth models.

The right kind of growth model is ongoing, inclusive and sustainable. This means ensuring that our cities are those where people can breathe, move and be productive, that the energy we generate comes from cleaner, cheaper sources and that our natural assets can continue providing the resources and environment on which depend the well-being of present and future generations. The chapter, 'The World We Are Yet to Build', looks at our cities—pollution, traffic congestion and the costs therein. The good news is we can course correct: first, with more compact, connected and coordinated cites; second, our energy systems and the positive steps the government has taken in this direction; third, bolstering private investment and finance to help bridge the existing infrastructure gap.

India is stepping up as a leader on the global stage on climate action and sustainable development. I look at some of the initiatives from the government and corporates and their positive actions giving examples of how businesses are embracing these causes. We are seeing unprecedented convergence around strong climate action from government and business and religious communities.

The hope around the Obama–Modi meeting was short-lived as the aspirations are unlikely to be fulfilled in a Trump regime. The G20 agenda may well be impacted too. However, the green economy is already one we understand in India. We have shown our global leadership along with France in the announcement of the International Solar Alliance. In the chapter, 'Rivals and Partners: China and India', I argue that going ahead, the success of both countries hinges on strong urbanization models and energy transitions. We can learn from each other's successes and failures. Our stories are already intertwined—China is the world's largest and cheapest supplier of solar and wind energy equipment.

In 'Building the BRICS of the Paris Agreement', I have highlighted how the BRICS countries (Brazil, Russia, India, China, South Africa) have already stepped up to reduce emissions and promote sustainable development. As a member of the BRICS Business Council from India, I am confident that the agenda around sustainability and green financing are central to India's BRICS engagement. Indeed, the first five projects of the BRICS-promoted New Development Bank (NDB) were all in the renewables space and its first debt issue was designed to finance clean energy. This chapter looks at the many positive steps taken by the BRICS member countries.

The chapter, 'The Case for a BRICS Rating Agency', looks at the important initiative of providing ratings on a broader emerging market scale, one which will offer sharper differentiation in credit ratings for the benefit of investors focused on emerging markets. Ideally, it should help reduce the cost of funds as projects and companies' ratings stand on their own and are not hampered by sovereign ceilings. For years, I was part of teams that suggested that credit rating agencies do not give such heavy weightage to per capita numbers, as when you look at emerging populous countries such as India and China, we inevitably look weaker than the scale and strength of our economies. There must be a different way of assessing projects in our countries. I am glad to see the idea of a new BRICS rating agency gather momentum as we look to build huge sustainable infrastructure at low cost.

WILDLIFE AND HABITAT CONSERVATION
The Wildlife Protection Act of 1972 established sanctuaries, national parks and tiger reserves which were over 660 in number till 2015. Forests are important carbon capture sites and the good news is that forest cover in some of our states is improving. By creating national reserve forests, we have protected our trees and biodiversity and also the waterbodies and rivers that form part of these areas. Most of this protection has come from Project Tiger where the fear of extinction of our national animal fortunately impelled us to protect our pristine forests.

Our family holidays are typically short trips into our jungles and my husband and I have attempted to capture some of our thoughts in the chapter, 'Wildlife and Habitat Conservation'. We look at tourism and how it has helped conservation and also discuss some ideas on how tourism can be enhanced with minimal negative impact on the environment. With appropriate training for naturalists, guides and drivers in our parks, we can improve the overall tourist experience beyond just chasing the tiger, encouraging interest in the varied bird life, fauna and foliage. We can support the forest department's effort in park management through better policing by providing many more trained eyes and ears and feedback. It is heartening to note how guides and drivers support a litter-free environment by picking up litter and also ensuring that visitors obey rules.

For a park to be successful, the communities around the forest need to benefit from tourism. The job opportunities are many, such as guides and drivers and provision of vehicles and their maintenance, employment in the lodges and local farmers providing food and vegetables. Villagers can also benefit from tourism, just like in Brazil and Africa, with exposure to their lives, culture and arts and crafts.

Poaching and trade in animal parts is known to fund terrorists and anti-national elements. We have looked at the challenges of containing poaching and believe there is a need to integrate efforts of the Forest Department with the police and paramilitary forces. The equipping of our forest guards and attracting new recruits is in dire need of attention. The job of a forest guard, despite being a government job, is not a career of choice.

Insufficient habitat is the biggest challenge—we are rapidly reaching a situation of overpopulation of tigers in our tiger reserves, of rhinos in Kaziranga (Assam) and lions in Gir (Gujarat). We can translocate animals to other national parks and also create new parks as we do have large tracts of forest that lend themselves to these animals and can be converted. However, we lack the administrative set-up and manpower to secure these areas from poachers and man–animal conflict and manage these forest tracts effectively. We need to reinstate the corridors that connected our forests, enabling animals to find new territories, helping their survival and strengthening their gene pools.

We have also put forward the idea of a wildlife coalition as a platform to bring all stakeholders together to share best practices, and collaborate and partner for improving the impact of their individual efforts. The coalition website can be the go-to point for these best practices and conversations around conservation, skilling, standards and administration. Researchers can meet and share their learnings. Through the coalition, priority funding requirements could be addressed. A wildlife coalition becomes a meeting point, both virtual and physical, for all players in the field as indeed we have seen at the ISC.

WOMEN

No book on these subjects is complete without looking at the role of women, as they are central to all programmes on sanitation, water, agriculture and forests. We need to empower them so that they assume leadership roles in implementation of such programmes in these areas and in bringing about behaviour change. The chapter, 'Women's Empowerment: Nurturing the Ecosystem', was prepared by me for NITI Aayog as inputs for the vision document on women, children and nutrition. It discusses the role of education and financial inclusion in enabling participation of women in the economy.

I have touched on the role of microfinance for building toilets or water harvesting. These loans are typically given to joint liability groups and self-help groups (SHGs, groups of seven to ten women coming together and standing guarantor to each other) and have a 99 per cent payment track record. However, providing loans is not enough. We need to train these women in livelihoods, business and financial management, so they can run microenterprises, earn money to repay loans and also save for the welfare of their families.

Government schemes like the Pradhan Mantri Jan-Dhan Yojana and Rajasthan's Bhamashah Yojana work on the premise that conditional and direct transfers have the highest impact on poverty reduction, particularly where the money is transferred to the bank accounts of the women in the family.

Lack of sanitation facilities leads to severe health problems and safety issues for women. Water scarcity also impacts women as the primary caregivers. Reports suggest that provision of water at the home liberates women from the chore of collecting and carrying water from a distance, freeing up their time to earn livelihoods and enabling girls to stay in school rather than be detained at home to collect water.

'Rural Indian Women: Sowing the Seeds of Social Transformation' suggests that the role of women in rural sectors, such as agriculture, forestry, animal husbandry, education and community projects, cannot be overlooked. The chapter looks at projects where women transformed their lands with the help of NGOs leading to social forestry, horticulture and orchards and implemented significant water harvesting and water storage facilities. Automation and use of latest technologies for irrigation to use water wisely may help women increase productivity and achieve even better results.

The social and economic advancement of India remain my predominant focus. If you have picked up this book, I am probably already speaking to the converted—you care about these issues. I have set out the agendas across the various topics covered in the book, indicating cross-linkages. I do hope they set you thinking, as the reader, if not agreeing with what I have written.

Water and Sanitation

Sustainable Sanitation: A Review

India, at present, is midway into the SBM. Since its inception on 2 October 2014, the MoUD and MDWS have been spearheading the programme with implementation happening at the State level. The key differentiator with SBM is the Prime Minister's ongoing focus which has percolated down to the district and block officials. It has also captured the imagination of the people of the country as several surveys show it to be the most appreciated of government programmes.

The SBM has been around for three years and has witnessed several notable achievements in reducing open defecation and mainstreaming sanitation, thanks to the focus on behaviour change, need-based capacity building and constant measuring of outcomes. This period has seen an increase from 42 per cent to 73.31 per cent of the national sanitation coverage. Seven states, 247 districts and almost 2.85 lakh villages have already been declared ODF. Nearly 22 per cent of the cities and towns have been declared ODF; 50 per cent of the urban wards have achieved 100 per cent door-to-door solid waste collection; and over 20,000 Swachhagrahi volunteers are working across ULBs while over a lakh are working in rural India. The number of schools having separate toilet facilities for girls has increased from 0.4 million (37 per cent) to almost 1 million (91 per cent). Bio toilets in railway coaches increased from 8,788 during 2016–17 to more than 19,000 by March 2017. In Rajasthan, water, sanitation and hygiene are integrated into the lessons of all children from classes III to VIII. All results and outcomes, in rural and urban India, are tracked through dynamic online systems. The Bal Swachhta Mission, envisioned to make children aware of the importance of hygiene and cleanliness at each Anganwadi centre, has been initiated. The 'Darwaza Band' campaign aimed at changing the behaviour of men has also been initiated.

The SBM has seen numerous analyses, discussions and conclusions being drawn about the programme. One of the recent media reports mentions that the government is not measuring ODF, and rather tracks funds spent on latrine construction while putting out numbers about sanitation. This is not entirely correct, as there have been efforts to measure ODF. Of course the modalities for the same can be debated and there may well be scope for improvement in the measurement protocols. Several sectoral experts are members of the Empowered Working Group (EWG), which is responsible for examining the survey methodology and setting protocols for the government's upcoming national survey through the Independent Verification Agent (IVA) under the World Bank project.

Since the launch of SBM, we have seen an increasing number of players willing to participate and contribute to the national sanitation mission. These include corporates, NGOs, multilaterals and media to name a few. Previously, we had seen national sanitation programmes like Central Rural Sanitation Programme (CRSP), Total Sanitation Campaign (TSC) and

Nirmal Bharat Abhiyan (NBA) evolving and contributing in improving the sanitation scenario. Those programmes focused on construction of toilets and achievements against targets. The SBM in addition brought a fresh perspective about the need to focus on behaviour change and equitable sanitation.

KEY DIFFERENTIATOR

One of the key differentiators of the SBM programme (and rightly so) is the decision by the government in November 2014 to make ODF the success parameter. It was made clear by the concerned ministries that progress will be tracked and evaluated only based on achievement of ODF. This caused a paradigm shift in the overall thinking of the implementers as ODF measurement has a direct relationship with behaviour change. This policy shift led to ODF Monitoring Committees (or Nigrani Samitis) being formed at village levels reflecting the community ownership of SBM. The monitoring committees' key tasks were not to count the number of toilets, but to ensure that no individual from the village resorts to open defecation. Anecdotal information and feedback from NGOs and others in the field suggest good progress on this front. The overall achievement is a reflection of the policy shift made by the Government of India in tracking the outcomes.

Sanitation, being a complex issue in a diverse country like India, encompasses a number of factors which are important determinants for the success of the mission. Sanitation has a direct relationship to caste, creed, religion and gender. A successful sanitation programme needs to address all such factors which make achievement of safe sanitation and ODF a very complex exercise. Additionally, India has a large number of disabled people whose sanitation needs also have to be met by providing customized solutions. Despite all these challenges, we have seen a marked improvement in the sanitation coverage in India since the launch of SBM.

However, achieving ODF status alone is not sufficient for the success of SBM. To make this programme a success, attention to the complete sanitation cycle is required, where toilets not only need to be built and used but the waste generated also needs to be collected and treated properly. The ISC advocates safe and sustainable sanitation, including design, implementation and practice, as is evident from its tag line—BUMT (Build, Use, Maintain and Treat), to complete the entire sanitation chain.

Achieving ODF is the collective responsibility of the entire nation and not only of the government. We have now reached a stage where the need for Behaviour Change Communication (BCC) has been recognized. One cannot deny that there is much to be achieved and that we 'have miles to go before we sleep'. We need to continuously and critically analyse our programmes with a view to course correcting and looking for practical solutions. We also need to recognize and appreciate all that has been achieved and replicate the successes.

Turning a large and populous country like India around is not an easy task. However, in less than three years, we see that India is already course correcting and with the momentum building, the pace of change, going forward, will be much faster.

Galvanizing the Power of Multiples

It has been more than three years since Prime Minister Narendra Modi raised the nation's consciousness from the ramparts of the Red Fort launching the SBM and highlighting the need to build toilets. The mission has succeeded in capturing the imagination of the masses, created awareness and brought cleanliness on the national agenda. While it inspires us all to work towards achieving the goals of sanitation with renewed vigour, it also challenges us as citizens of the country, individually as well as jointly, to work towards delivering the goals set by the government to be ODF by 2019. Various government departments, down to the Gram Panchayat, are working towards this, creating a ripple effect. To bring about a paradigm shift and create a wave of behavioural change around sanitation, a lot needs to be done in the future through a joint effort of all stakeholders.

For most corporates, their core business is not sanitation, despite which, they are making tremendous efforts through their CSR wings and foundations to support the SBM. Given the large business opportunity with a target of 110 million toilets to be constructed, it would be good to see some large, credible corporate players/other organizations emerge to whom the rest of the corporate community could turn to for advice and outsource their needs. Here of course it needs to be ensured that not only are quality toilets constructed, but used, cleaned and maintained. A massive BCC is urgently required to address open defecation. This provides a huge employment opportunity, and providing skills to people who would be able to use these skills going forward with the large opportunities from initiatives such as Smart Cities, etc. Waste disposal also needs to be addressed and done properly. This can also provide large business opportunities and employment generation.

Over the years, in many parts of the country, silent revolutions have been going on, transforming people's perception on sanitation while making more toilets available concomitant with behavioural change. The transformation of seven states, 247 districts and almost 2,85,000 villages demonstrates that sanitation for all can become a reality and success can be achieved through a collective effort. It will be critical to ensure that usage and community engagement continues. Examples of community toilets working well in the slums of Dharavi and community-led approaches started in Nadia (West Bengal) and Bikaner (Rajasthan) and later followed by others need to be replicated. Similar good practices and technological innovations exist in the sanitation sector across the country. Technological innovations have been implemented in both rural and urban areas. Such local solutions need to be scaled up and technologies upgraded to have an impact of the magnitude that India needs going ahead. Customized solutions for disabled people and senior citizens also need to be adopted. Efforts to create a framework and enabling environment for including policies, strategies and protocols and the conditions for the fulfilment of the need for universal sanitation and hygiene are critical—for women and men, children, adolescent girls, people with disabilities and the elderly. To address sanitation needs of disabled people, a handbook on different technological options, and Menstrual Health Management (MHM) guidelines have been released by the MDWS.

Developing and promoting standards for toilet construction is another area of prime focus to ensure long-term sustainable asset creation. India needs a cadre of masons, plumbers, architects and sanitation experts without whom there will be a huge lacuna in the delivery mechanism and a backlash if poorly implemented toilets fail and fall to disuse.

The sanitation space in India presents some unique challenges intertwined with the complexities of caste and cultural practices. The preference in large geographies for open defecation despite fully functional toilets being available in the vicinity highlights the complexities of age-old traditions deeply rooted in the cultural milieu of India. The differential impact of poor sanitation practices and infrastructure on maternal and child health and safety is well-documented. Repeated diarrhoea caused by open defecation leads to 'stunting' which affects the child's physical and mental development, affecting the productivity of our people. Unfortunately, there is almost a complete lack of awareness of the health aspects of open defecation. The exceptions to the rule are touted as: 'My grandfather who is over 80 years has been going to the fields since his childhood, and he's fine.' Also quotes like. 'How can you have a toilet around the house when cooking is also done inside the house?' Also missing from the popular discourse is the need to build an ecosystem for sanitation that involves market creation for service provision, products, technologies and institutional capacities for standards and certification. Matching the demands with supply will be critical to address the sanitation value chain. After all, the entire advocacy around behavioural change needs to be strongly backed up with a robust ecosystem that can take care of the BUMT value chain.

From the time the SBM captured the nation's imagination, a lot has been written and said about what a practical strategy for India must look like. While there is a consensus on the areas of action, there is need for more collaboration and coordination between stakeholders. This would increase efficiency, prevent reinventing the wheel, and therefore, increase the impact. The presence of a coherent force to tie the loose threads together and connect the dots is the need of the hour. The power of multiples that can be created by galvanizing the joint force of government, corporates, civil society, practitioners, media, communities and societies at large, cannot be overstated. This is what the ISC (inaugurated on 25 June 2015 by Birender Singh, then the Minister for Drinking Water and Sanitation, Government of India) aims to do, as a catalyst, facilitator and galvanizer.

The concept of the ISC was initiated around the end of 2013 by a group of people from different organizations. The Coalition has been established to bring organizations and individuals together on a common platform to find sustainable solutions for sanitation. It will bring momentum and create a strong platform for aggregation of knowledge and networks with nationwide outreach, focusing on models for achieving sustainable sanitation in alignment with the SBM and its goals. The very philosophy of the ISC embodies the principles of BUMT, signifying a holistic approach to sanitation rather than a mere emphasis on toilet construction. The Coalition endeavours to institutionalize this integrated approach in the way sanitation is perceived in the country. The four task forces of the Coalition, viz. Advocacy, Branding, and Communication; Identification and Dissemination of Best Practices; Engagement with Central and State Governments; and Partnerships and Collaboration, have been established with cross-sector representation. These task forces are working closely with stakeholders in building partnerships, documenting and disseminating best practices, creating peer-to-peer influence and advocacy. Community building and skill development are the other areas of focus.

The Coalition works with the MDWS on an annual National Conclave, engaging industry chambers and corporates. State-level engagements with the governments of Rajasthan and Maharashtra are under way. The Coalition has initiated dialogue on skill development in schools and has brought microfinance institutions (MFIs) into states with its partner, Water.org, to provide finance for sanitation. The Coalition has also partnered with the SuSanA to initiate thematic discussion forums in India. The growing number of partners joining the Coalition is encouraging.

Success must be achieved through a community-led process with local leaders, such as panchayats and other community groups, supported by political and bureaucratic leaders. Change must be people-centric and designed with the end user in mind. Community Led Total Sanitation (CLTS), mainly in rural areas, is one such proven community-led process whose effective implementation ensures long-term success. The CLTS involves facilitating a process to inspire and empower rural communities to stop open defecation and build and use latrines. It uses participatory methodologies to develop awareness of the risks of open defecation and facilitate the community with a self-analysis of their health and sanitation status. Its aim is to 'ignite' communities to cease open defecation and commence toilet construction using local materials. The CLTS has been recognized by the UN as one of the most effective approaches to promoting sanitation and achieving the

Millennium Development Goals (MDGs) for sanitation. Global Citizen India is one such sterling example of creating a strong citizen engagement platform, and particularly working with youth communities—since the future leaders and champions of social change are the young and India must leverage its great demographic edge of a vibrant youth! Their strength in numbers and their energy and receptivity to new ideas should be harnessed for creating a strong on-ground movement. The ISC has partnered with Global Citizen India, a social action platform that comprises a distinctive mix of events, grassroots activism, media campaigning and online activation, to catalyze India's fifteen-year journey towards achieving the SDGs, and to bring about the end of extreme poverty. We need more of such activities and platforms which believe that sanitation is the key to tackling extreme poverty.

Coalitions have the power to create, enable and bring a momentum to endeavours of significant proportions and scale. The power of coming together as one is far greater than the power of doing things alone. The SBM has presented us with a historic opportunity and the formation of this coalition is a timely initiative that underscores the importance India assigns to sanitation and cleanliness. Ultimately, it is each of our responsibility to move the sanitation agenda of the country forward. Whether it is stopping someone from littering, engaging with our local communities or politicians on the issue or providing adequate facilities to people in our own workspaces or homes, a small step can go a long way. Only collectively, through the creation of a people's sanitation movement, can we achieve a truly Swachh India.

We believe that we can together make a huge difference.

Corporates as Partners in Achieving a Swachh Bharat

The magnitude of the sanitation crisis in India cannot be overstated, given its significant social, economic and environmental repercussions. It is one of India's greatest challenges. Of the 1.1 billion people in the world who defecate in the open, more than half reside in India alone.[1] The harmful effects of inadequate sanitation in India continue to have a severe impact on a healthy population, education, a productive workforce, economic growth and our path towards becoming a truly developed country. Globally, 88 per cent of diarrhoeal deaths are due to lack of access to proper sanitation facilities.[2] In India, diarrhoeal diseases account for 1,600 deaths daily—the same as eight jumbo jets crashing every day.[3]

INDIA'S $53.8 BILLION PROBLEM

Due to adverse economic impacts and costs of inadequate

---

[1] 'Squatting Rights—Access to Toilets in Urban India', a report by Dasra and Forbes Marshall, http:// www.dasra.org/pdf/SquattingRights_Report.pdf
[2] 'Safer water, better health: Costs, benefits and sustainability of interventions to protect and promote health', World Health Organization
[3] Water.org

sanitation, 6.4 per cent of India's GDP and 73 million working days are lost. Further, many households remain unconnected to the sewage system, with over 1.3 lakh tonnes of human waste generated every day and this number is ever increasing.[4] Furthermore, an estimate of over 73 per cent of all faecal sludge generated in urban India is left untreated in the environment.[5] Collectively, that means India's sanitation crisis is a $53.8 billion (₹2.4 trillion) problem. To put things in perspective, the annual expenditure budget for the fiscal year 2016–17 of the central government was just a little over $300 billion.

Given that sanitation in the Indian context is multifaceted, layered in behavioural, social and cultural complexities, even within the country there is a wide disparity. While 85.9 per cent people in Odisha and 82.4 per cent in Bihar do not have access to toilets, seven states have been declared 100 per cent ODF. Against the context of these statistics, while Goal 7 of the MDGs went unmet, adequate sanitation has warranted its own place as Goal 6 of the SDGs. With the current state of the sector and the mammoth task at hand, the government's target to make India ODF by 2019 necessitates the involvement of, and collaboration amongst, multiple stakeholders, including the corporate sector.

In order to achieve the required speed, scale and sustainability, India must leverage the strengths of its private sector as well as address the often unrecognized and misconstrued sanitation problem in urban spaces. Moving ahead, this will be critical given India's rapidly growing urban areas with a burgeoning urban population in dire need of adequate sanitation provisions.

---

[4] https://docs.gatesfoundation.org/Documents/ICO_Letter_ENGLISH.pdf
[5] Central Pollution Control Board, 2015

Historically, the onus of providing solutions for societal problems has been seen solely as a government prerogative. To provide the necessary impetus and fill the vacuum created by limited public resources, the non-government sector has evolved, which also brings in the much-needed accountability of public spending. The private sector and corporates, driven by the motive of monetary profits have traditionally been viewed as outsiders in this scenario. Their efforts were seen to be limited to their own 'catchment areas', i.e. areas in and around their plants and factories. If seen through the lens of the CSR mandate, this approach makes perfect sense. This creates co-benefits or shared value which benefits the company by creating goodwill and a competitive advantage while accruing benefits to society. In addition to leveraging business resources to create social value, CSR also provides the opportunity to unlock newer areas of collaborations with local government and institutions.

ILLUSORY STATISTICS AND INDIA'S MISCONSTRUED URBAN SANITATION PROBLEM

In urban India, the extent of the sanitation problem is often misunderstood. When viewed in totality, data suggests that rural India is far worse off than urban India in terms of adequate sanitation coverage and thus requires more attention and investment from multiple stakeholders, including corporates. For example, according to the 2011 Census, access to improved sanitation for households in rural India was a mere 31 per cent vis-à-vis 81.4 per cent in urban India. Therefore, according to this, only 18.6 per cent of urban households are reported to have no toilets. Moreover, in urban areas, the percentage of houses with improved toilets having water closets has increased from 46.1 per cent in 2001 to 72.6 per cent in 2011, while correspondingly the percentage of houses with pit toilets has decreased from 14.6 per cent to 7.1 per cent.[6] Thus, the situation in terms of availability of physical infrastructure in urban areas seems to be far better off than that in India's rural areas. However, when the same data is bifurcated particularly taking into consideration the urban poor population, a different picture emerges. Here, it is important to acknowledge that the 2011 Census states that nationally, one in six Indians lives in an urban slum, the hub of India's urban poor.

Two variables play a critical role when talking about urban sanitation—availability of space and population density. As outlined above, inadequate sanitation is often compounded in urban slums. Here, close living proximity and crowded conditions, open drainage and lack of maintenance around existing facilities contribute to the outbreak and spreading of diseases, causing significant health risks for these vulnerable populations. Even for those who have access to improved sources of sanitation, the quality of these services remains inadequate and unequally divided amongst the population.[7] As India confronts rapidly expanding slum populations, and with over 50 million people forced to defecate in the open[8], the problem and associated health risks will continue to grow rapidly if left unchecked.

WASTE MANAGEMENT NEEDS URGENT ATTENTION

In the urban context, a particularly important component that requires significant attention is the issue of solid and liquid waste management (SLWM). Managing human waste safely

---

[6] http://www.censusindia.gov.in/2011-Common/CensusData2011.html
[7] WHO/UNICEF Joint Monitoring Program (JMP) study on Water Supply and Sanitation, 2015
[8] 'Squatting Rights—Access to Toilets in Urban India', a report by Dasra and Forbes Marshall, http:// www.dasra.org/pdf/SquattingRights_Report.pdf

adheres to a mechanism whereby human waste does not come in contact with humans and the environment, and is disposed of safely. The benefits of sanitation accrue most if all have access to good-quality toilets and the entire waste is treated properly. If there is no universal access to toilets and 100 per cent treatment of waste water, then the entire public faces increased incidences of waterborne diseases, vector-borne maladies, higher water treatment costs for municipalities and individual households and a filthy environment. In urban spaces with a higher density of population, this becomes even more complex as there is often extreme pressure on the infrastructure due to over-utilization and ill-maintenance.

Even where access to sanitation is available, many urban residents use toilets that are not connected to underground sewerage networks. It is estimated that 75 to 80 per cent of water pollution by volume is from domestic sewerage. In a rating exercise undertaken by the MoUD in 2010, it was observed that none of the 423 cities that were rated were found to be healthy and clean. In fact, around 190 cities were reported to be on the verge of an environmental crisis. Further, faecal sludge is the human waste from on-site sanitation (OSS, that is, systems below the ground, not connected to sewers). In India, nearly 1,200 cities have fully OSS systems. And even where treatment facilities exist, 40 per cent do not comply with the basic standards. These jarring statistics are further illustrated by Figure 1. Furthermore, it is also important to mention that manual scavenging, although prohibited, continues to take place across India. A manual scavenger is defined as 'a person engaged in or employed for manually carrying human excreta'[9], and is related to the unsafe and undignified emptying of toilets and

---

[9]https://thewire.in/54596/what-is-manual-scavenging/

pits, as well as handling of raw, untreated human faeces. Both the issue and the dire lack of data around the same must be addressed.

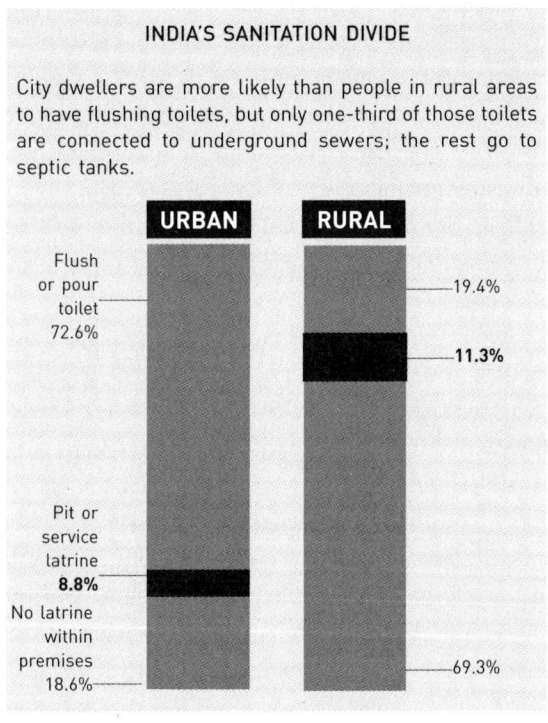

Figure 1

Source: Sunita Narain, 2012, www.nature.com

Today, while there is an emerging recognition of the importance of septage management, a lot still needs to be done in terms of advocacy, planning, execution and the ability to create sustainable business models around the same.

In rural areas, examples of solid waste include waste from kitchens, gardens, cattle sheds, agriculture and materials such as metal, paper, plastic, cloth and so on. They are organic and inorganic materials with no remaining economic value to the owner produced by households and commercial and industrial establishments. Most household waste in rural areas is organic, with little inorganic material, and is non-toxic. Because of its environment-friendliness, composting is a highly suitable method for waste management in rural areas. Similarly for liquid waste, industrial waste water is generated by manufacturing process and is difficult to treat. Domestic waste water in rural areas includes water discharged from homes and educational institutions like schools and Anganwadis. Estimates show that people in rural India generate 15,000 to 18,000 million litres of liquid waste (grey water) and 0.30 to 0.4 million metric tonnes of organic or recyclable solid waste per day.[10]

The objective of waste management in rural areas should be to promote a clean and healthy community where all the waste is treated and disposed of safely. In rural areas, management of solid or liquid waste is much easier than urban areas, due to various factors like organic nature of most of the waste generated, absence of highly contaminated industrial wastes, various options for treatment with no space constraints, etc.

The SBM has been designed to encourage participatory approach from communities at grassroots levels. Community or Gram Panchayat-level functionaries are responsible for design, implementation, operation and maintenance (O&M) of SLWM systems, with support from respective state governments. There are mechanisms to tap suitable third parties to undertake various

---

[10]'Solid & Liquid Waste Management In Rural Areas: A Technical Note', UNICEF

construction and management aspects under the supervision of the gram panchayats and communities. In such cases, it is imperative that there is absolute clarity in the roles and responsibilities of various stakeholders in managing SLWM systems. Community contribution and appropriate user charges for sustainable SLWM initiatives are also essential.

THE SBM IMPERATIVE

The engagement of corporates in the WASH sector in India, specifically in sanitation sector, is not a new phenomenon. However, what is unprecedented today is the renewed buzz and energy around which corporates have been called to action. Spurred on by the SBM, which was launched by the Prime Minister of India with a clarion call to address the deep-rooted issues of sanitation which have serious economic and health impacts, the discussion around corporate engagement is no longer limited to a particular type of organization confined to one-time financial contributions.

This Big Hairy Audacious Goal (BHAG) of the development sector aims to accelerate the coverage of universal sanitation by focusing on community-led models. It differs from its predecessor schemes as it aims to create a sanitation movement in the nation. This can only be possible when multiple players come together to collaborate and leverage each other's strengths.

The government on its part has put the much-needed spotlight on sanitation and actively encourages corporate participation. In letter and spirit, it is important to acknowledge that the call to action to corporates, particularly from the government, envisages support beyond just the funding or construction of toilets. In the recently issued guidelines of MDWS, launched first using the platform of the ISC, the document asserts that 'the creativity and efficiency of the corporate sector, and their management and financial resources can help in achieving the vision of a Swachh Bharat'.[11] Thus the modalities of this support may be as basic as mere financial assistance, by way of the Swachh Bharat Kosh, or getting involved through their technical expertise, marketing excellence and outreach support. The government expects the corporates to engage through a constructive value-driven approach, rather than a hollow compliance-driven approach to ascertain greater stake, goodwill and brand equity.

It, therefore, comes as no surprise that today many corporates are responding enthusiastically to this call to action, with a majority leveraging Section 135 of the Companies Act, 2013, with a view to complying with the CSR responsibilities set out therein.

Realizing the importance of fruitful corporate engagement, the government has tried to provide a structure by releasing a private sector engagement framework.

It is imperative to highlight specific activities in sanitation where the private sector can support the government (both central and state). With their expertise in the areas of innovation in technology, project management and scalability, viewing the role of corporates as mere funders would be detrimental to the mission of achieving sustainable sanitation.

REFLECTING ON CURRENT TRENDS

With the aim to capture CSR trends in the sector, the ISC released a report in 2016 titled, 'CSR in Water, Sanitation and Hygiene (WASH): What are India's top companies up to',[12] anchored by Samhita Social Services, a partner of the Coalition,

---

[11]Ministry of Drinking Water and Sanitation, 2016
[12]http://www.samhita.org/csr-in-wash-what-are-indias-top-companies-up-to/

analysing the top 100 companies with the largest CSR budgets. The report found that 90 per cent of the companies have at least one CSR programme in WASH. The break-up of the nature of companies is illustrated in Figure 2. However, the findings demonstrated that 75 per cent of the companies were supporting programmes related to infrastructure creation—construction of toilets and water facilities, with limited attention to behaviour change programmes. It was only a handful of companies that were engaging across the value chain of sanitation that includes all the components of BUMT, a concept with sustainability at its very core. Some companies are also implementing O&M programmes. Additionally, it was also found that industries with a strategic interest, like Heavy Engineering & Manufacturing and Fast-moving Consumer Goods (FMCG) companies were more likely to support WASH programmes than other industries. An important finding of the report was that most of these WASH programmes are concentrated in rural areas.

To reimagine the corporate as a partner, it is important to first understand why many have fallen into this compliance, number-driven approach. The primary reason is that they have not been integrated into the broader sanitation ecosystem. The focus on supply-side deliverables stems from the long-standing understanding of sanitation programmes as mere toilet construction activities. Even when there is intent to move beyond this conventional domain, the lack of required tools, knowledge base and hand-holding support severely limits designing and implementation of holistic interventions. Further, given the government's desire to reach various supply-side infrastructure targets, the SBM also runs the risk of a narrow agenda, with many corporates viewing it as a 'toilet-building' programme. However, this number-driven and disconnected approach can be disruptive to the broader objective to end open defecation. As Arghyam, a partner of the Coalition, has outlined, 'Sanitation has to be invested in holistically or it can do more harm than good. An investment in a toilet has to be preceded and succeeded by pro-active and consistent efforts at behaviour change to ensure continued usage by all members.'

Reflecting the overall nature of the sample, the Heavy Engineering & Manufacturing industry dominates the number of companies and CSR programs in WASH

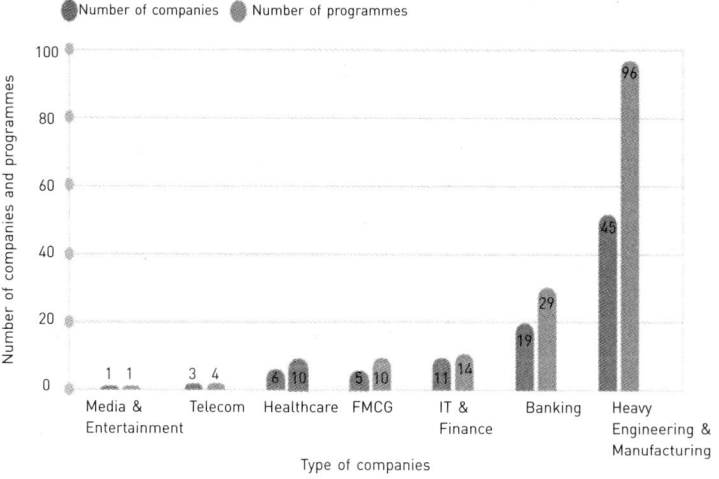

90 companies supported programmes in Water, Sanitation and Hygiene (WASH)

Figure 2

As the report concludes, 'For a well-functioning WASH market, gaps in the ecosystem need to be addressed. More attention needs to be given to financing solutions for stakcholders, innovations should be encouraged across the value chain, actionable knowledge produced and disseminated and stakeholders supported through capacity building initiatives.'

CORPORATES AND THE URBAN-RURAL DIVIDE

According to the report's findings, eighty-six of the 100 surveyed companies published information on geographical coverage. Of that, 52 per cent were focused exclusively on rural areas, compared to only 17 per cent which focused on urban areas. The remaining 31 per cent's efforts were spread across mixed geographies. This is illustrated in Figure 3.

Corporate Engagement in WASH in rural vs urban areas

		Urban coverage		
		No	Yes	Total
Rural coverage	No	–	17%	17%
	Yes	52%	31%	83%
	Total	52%	48%	100%

Figure 3

The clear preference of corporates to engage in rural areas can be based on a series of assumptions, including stakeholder interest in terms of community engagement around factories, the availability of space for construction and the ease to navigate rural leadership structures like the Panchayat. Bigger geography in rural areas is also one of the reasons.

A deep dive into the data of those corporates, both the 17 and 31 per cent that indicated their engagement in urban WASH, shows that a majority of companies worked with schools and construction in municipalities while others also engaged in peripheral activities like Swachhata Saptahs (cleanliness drives). The Swachhata Saptahs also took place mostly in the vicinities of respective corporate headquarters and were often seen as a branding exercise in which top management could also participate. Other awareness drives included those centred on

Information, Education and Communication (IEC) and BCC activities, RO/water purifier plants and biodigesters for slums and community households.

What is apparent is that there is both a pressing need and opportunity for corporates to engage in urban sanitation. At present, a majority continue to focus their efforts in rural areas. In order to shift towards increased corporate engagement in India's urban sanitation problem, it is important to recognize both the challenges and opportunities that exist for the same.

A CHALLENGE AND AN OPPORTUNITY

Corporate engagement in urban sanitation has many challenges associated with it. Across various testimonials and reports, multiple reasons have been cited for this, spanning technical, operational and administrative issues in conjunction with limited space availability and high population density in urban spaces. Furthermore, some corporates and NGOs cited the difficulties associated with the red tape working culture of the ULB management and thus the complexities of navigating through complicated urban organizational structures. The small number of corporates engaged in urban sanitation highlighted certain barriers, including the lack of usable knowledge on best practices and scalable models in the urban space as well as lack of understanding of the problem based on available data, the inability or perception of the inability to find the right implementation partners, the difficulty to quantify impact and difficulty in traversing government networks.

However, while these challenges exist, there remains both a huge need and an opportunity for corporates to enter into the urban sanitation space. Just the required capital expenditure for the SBM urban (U) programme is ₹1,31,137 crore.[13] SBM (U) covers 4,041 statutory towns. The overall target of the mission is to construct 1.04 crore units of individual household toilets, and 5.08 lakh units of community and public toilets in urban areas. To improve citizens' access to sanitation, seven key mission objectives have been identified. These include:

1. elimination of open defecation
2. eradication of manual scavenging
3. modern and scientific municipal solid waste management
4. to effect behavioural change regarding healthy sanitation practices
5. generate awareness about sanitation and its linkage with public health
6. capacity augmentation for ULBs
7. to create an enabling environment for private sector participation in CAPEX (capital expenditure) and OPEX (operating expenses).[14]

To complement the funds earmarked by both the Centre and state, the government is looking for the balance funds to be generated through other sources in the form of beneficiary contribution, user charges, CSR funds and private sector participation, among others. Additionally, the guidelines point towards how the government is also looking towards the private sector for support with public toilets.[15] Undoubtedly, the sheer scale and envisioned pace of SBM (U) both necessitates and is

---

[13]Centre for Policy Research and Confederation of Indian Industry. 'Swachh Bharat: Industry Engagement – Scope & Examples'. <http://www.cprindia.org/sites/default/files/policy-briefs/Swachh%20Bharat-Industry%20Engagement%20Report.pdf>

[14]Ministry of Urban Development, 2014, and Ministry of Drinking Water and Sanitation, 2014

[15]Ministry of Urban Development, 2014

an opportunity in itself for corporate sector engagement. What is required then is the creation of an enabling ecosystem to encourage and support the same.

In fact, the government has put into place many avenues to ease the engagement of corporates with SBM, including the creation of a Swachh Bharat Kosh and the recently launched Swachh, Swachhata Augmentation through Corporate Helping Hands portal. The Swachh Bharat Kosh is a special corpus set up by the government two years ago to mobilize funds for the SBM. It expects to attract funds from potential donors comprising public and private companies in addition to philanthropists. It is a crowdfunding platform to spearhead the engagement of private sector in city-level initiatives of the SBM.

Since its launch in September 2014, the Swachh Bharat Kosh has thus far received more than 600 crore including interest as detailed in Figure 4. Of this, the Union finance ministry that administers the fund has already sanctioned ₹382 crore to different states for implementing sanitation projects. Official data reveals that the fund's top donors have also remained largely the same in the last one year. Mata Amritanandamayi Math, with a contribution of ₹100 crore, remains the largest donor, followed by Larsen & Toubro ([L&T]; ₹60 crore) and Rural Electrification Corporation (₹25 crore). Other top contributors include Indian Railway Finance Corporation, IFFCO, ITC Ltd and Nuclear Power Corporation, who have each donated about ₹10 crore. According to the operational guidelines for the fund, the donations will be used for 'improving cleanliness levels in rural and urban areas, including in schools' through activities like the construction and repair of toilets, and providing water supply to the toilets.[16]

---

[16]*The Hindu Business Line*, RTI, March 2016 and *Hindustan Times*, July 2016

Source: Hindustan Times, July 2016

Figure 4

Many corporates have expressed hesitation around providing large sum donations into a single corpus and instead prefer to work on projects with direct accountability and project management. In order to align with this thinking, the MoUD has ensured that the Swachh portal works as an interface for collaboration between ULBs and corporates along with willing entrepreneurs along the sanitation value chain. The platform aims to connect city municipal commissioners with private individuals and companies who are interested in funding and getting involved in city-level SBM projects of building toilets and solid waste management infrastructure/services.[17] The platform has listed its objectives as: to encourage private sector participation in ULB initiatives of SBM; to attract private sector capital and expertise to bridge

---

[17] https://swachh.org.in/about-us.htm

gaps in funding of SBM; to expedite SBM targets and provide a transparent monitoring mechanism; and to provide a hassle-free process for corporates to choose, invest in and monitor the projects.[18] The platform presents a series of projects mentioning types, sub-types and cost and duration, as shown in Figure 5, for corporates to pick up along with status of work for the existing projects and project reports. The projects could be of interest to corporations as potential CSR interventions or even for smaller entrepreneurs as a for-profit activity.

Figure 5

Source: www.swachh.org

---

[18]https://swachh.org.in/about-us.htm

Thus, the government has taken steps to create tools in order to ease corporate engagement in sanitation. As outlined above, under SBM (U), some initiatives are aimed specifically at urban spaces, inviting corporates to enter into partnership frameworks with city-based governing bodies. However, while these facilitatory tools have been set up, corporate engagement in the space of urban sanitation falls short of the need.

CORPORATE GOOD PRACTICES

With the mandate of bringing these good practices to the fore, the ISC has compiled some corporate good practices. These can be broadly classified as business interest (where business objectives are combined with social development), stakeholder interest (where companies are engaged in the community supply chain), companies with CSR interest (social development that may involve a community focus) and catalyzers (with competencies that can spur cross-cutting impact at scale—media, technology companies). They highlight the wide range of work done by companies across the entire sanitation spectrum right from creation of innovative infrastructure (Havells) to skill development (Kohler, Reckitt Benckiser [RB]), behaviour change advocacy (IL&FS, HUL), community engagement (Cairn), etc. There are many companies from the private and public sector which, by their approach, aim to change the corporate engagement in WASH.

For example, a corporate volunteering programme is under way through the ISC's Swachhata Doots programme. This programme is highlighted in a later chapter in this book, but in brief, it is based on a programme conceived by HUL for reaching out to villages with the message of 'Swachh Aadat' (Clean Habit). Each factory works out a congenial schedule that enables volunteering factory workers to set aside some time to act as Swachhata Doots. The programme creatively bundles various facets to create a strong employee volunteering programme. The facets are: a large network of factories and factory workers located in remote locations; an innovation in media to make the message reach far and wide; and the will and desire to contribute to the nation. The ISC, with Dalberg and HUL's support, is helping implement this programme for other corporates as it works to create a national sanitation movement.

Other examples of corporate engagement by ISC partners are:

HUL: HUL's Suvidha Centre is a purpose-built, sustainable community centre which addresses the hygiene needs of low-income urban households who face severe challenges due to lack of infrastructure and facilities. The Centre—developed in close partnership with Mumbai Municipal Corporation and Pratha, a non-profit community-based organization—provides toilets that flush, hand-washing facilities with soap, clean showers, safe drinking water and state-of-the-art laundry operations at an affordable cost. The Centre promotes a safe and welcoming environment for everyone. Suvidha was devised, developed and built in close consultation with the local community. It is a community centre that meets real needs. What makes it more unique is that it is a market-based solution that is designed to be affordable and replicable. The ISC is encouraging other corporates to adopt this model, and corporates such as HSBC have signed up.

Mahindra & Mahindra: Mahindra & Mahindra is actively involved in carrying forward Swachh Vidyalaya Swachh Bharat Abhiyan. The Mahindra approach to the sanitation problem in schools is unique. Its focus is more on girls' schools and schools in rural areas. The design mostly consists of toilets in a block configuration model, usually comprising five individual units. The designs have carefully integrated special needs of children, including needs of those differently abled. The toilet blocks include eco-friendly waste water disposal and efficient use of water in the toilets. They are also equipped with graphics and visuals illustrating key hygiene practices and measures. The model takes special care to find customized solutions that take into account the requirements of the location with maximum utilization of locally available human and material resources. This tailored solution has successfully generated public participation and interest.

Bharti Enterprises: Bharti Foundation, the philanthropic arm of Bharti Enterprises, launched the Satya Bharti Abhiyan, a rural sanitation initiative in 2014. Working closely with the district administration, in its first phase over 18,000 toilets were constructed benefitting over 87,000 individuals across Punjab's Ludhiana district. In November 2016, Ludhiana became the second self-declared ODF district of Punjab. Recently, the Foundation has signed a Memorandum of Understanding (MoU) with the Punjab government to jointly provide over 50,000 toilets for rural households in Amritsar district.

Reckitt Benckiser: RB runs the ambitious programme, Dettol Banega Swachh India campaign to address the sanitation and hygiene crisis in India, investing ₹100 crore over five years. The campaign is based on these key pillars: driving behaviour change towards hygiene and sanitation practices; ensuring best-in-class, on-ground execution and using RB India's expertise in hygiene related products to improve the state of sanitation in the country. RB, in partnership with U.S. Agency for International Development (USAID) and Ernst & Young (EY), launched the highly innovative Hygiene Index programme. The Index will be calculated for more than 100 cities based on primary and secondary data collection and stakeholder consultations. Using the tool, the performance of the cities will be benchmarked and easily comparable, along with analysis of the best practices adopted.

Larsen & Toubro: Technology, engineering, construction and manufacturing giant L&T has taken up integrated community development projects as part of CSR activities in severely water-stressed areas spread over three states with a view to improving quality of life. Sanitation interventions are designed from a holistic perspective focusing not just on toilet construction but also on integrated community development projects including awareness drives, mason trainings, setting up monitoring committees, etc. In one of the intervention areas—Bhim block, Rajsamand district, Rajasthan—one hamlet of Jalpa revenue village under Kookara Gram Panchayat has already been declared ODF. L&T is also one of the corporates to have contributed to the Swachh Bharat Kosh.

HSBC: HSBC has established comprehensive CSR initiatives in the sanitation space. This includes school sanitation, household sanitation and community-level sanitation programmes in urban and rural areas. These programmes have interventions that go beyond providing toilets, and function all along the WASH value chain. They partner with thirteen WASH non-profits with a strong track record and proven capability to solve complex issues. Their partners across Gurgaon, Delhi, Rajasthan and Maharashtra are implementing a range of models that cover behaviour change, policy, provision of basic services and financing, along with infrastructure building, access to water and solid waste management. Partner organizations working with HSBC are encouraged to work in conjunction with government to leverage all possible resources and ensure that programmes are carried out efficiently and effectively. HSBC investments across 2015 and 2016 in WASH have enabled access to safe water for over 92,000 people, provided improved sanitation to over 52,000 people, and are projected to benefit another 162,000 people over the next three years.

Public Sector Undertakings (PSUs): The big push to the SBM has come from the PSUs. The National Thermal Power Corporation (NTPC) was the highest spender on CSR activities earmarking ₹491.8 crore in 2015–16, and allocating a substantial ₹285 crore of this amount for construction of toilets. For the Swachh Iconic Places, another flagship activity under the SBM, the public sector has been the major contributor. The goal of the initiative is to improve the cleanliness conditions at these places to a distinctly higher level. All the iconic sites have designated PSUs for financial and technical support.

MOVING THE NEEDLE: THE NEED TO CREATE A SUPPORTIVE ECOSYSTEM

To move the needle and both invite and attract more corporates to engage in the urban sanitation sector, corporates must be exposed to and understand both the challenges and opportunities at hand. In fact, the lack of a supportive ecosystem for collaboration remains a hindrance for corporates to move towards the role of a partner, even though there are many corporates, from marketing agencies to manufacturing companies, who are willing and able to engage in the sector. Many development partners also conventionally follow certain partnership stereotypes attributing corporates as mere funding sources for projects. To move forward, corporates must be integrated into a transparent and permeable sanitation ecosystem to enable partnerships, knowledge sharing, capacity building initiatives and a platform for communication and exchange. This includes incorporating SMEs into the mix, and in turn encouraging Public–Private Partnership (PPP) models that encourage innovation of affordable yet aspirational products. Further, to maximize the collective contributions, we must tap into the tremendous potential in creating entry points for corporates across the value chain of BUMT. Entrepreneurship across the sanitation value chain will play a critical role in converting those who openly defecate to practice safe sanitation. As there is a pressing need to encourage innovation across the life cycle of a sanitation programme, we need to encourage more entrepreneurs in these activities, including looking at profitable distributed models for waste management. Therefore, corporates are essential not only for investing through CSR, effective execution and scaling up, but also for creating innovative technologies and service models for delivery of services.

The ISC was formed with the vision to enable and support such an ecosystem, seeking to be an aggregator of knowledge and networks with nationwide outreach, focusing on models for achieving sustainable sanitation in alignment with SBM and its goals. While the Coalition through its network of partners addresses both rural and urban sanitation, it recognizes the importance of urban sanitation and the role that corporates can play to achieve the required impact and scale. In particular, the Coalition advocates that it is important to view the corporate not just as a funder but instead as a partner. When thinking of the role of corporate engagement in the sanitation space, particularly urban, we must base this on the value additions they can both bring and acquire. This could range from last-mile connectivity in terms of advocacy, skill development and capacity building, ecosystem building including provision of products and services, to a natural progression of their own businesses and aligning that with CSR and ensuring the shift towards 'sustainable' models for sanitation.

At the ISC, we have come across examples of corporates who are actively working in this sector through five primary means. First, those companies with a business interest including provision of products and services combined with social development, such as FMCG companies. Second, those with a stakeholder interest, wherein companies engage in the community supply chain. Third, those with a CSR interest that invest in social development. Fourth, those that act as catalyzers with competencies that can spur cross-cutting impact at scale, such as media and technology companies. And fifth, those that engage in volunteering with a focus to create a company culture of caring beyond business focus, driving loyalty and satisfaction amongst employees.

By creating a supportive ecosystem, such examples can be shared widely and other corporates can also be encouraged to enter into the sanitation space and create partnerships across stakeholders and the value chain.

THE WAY FORWARD

Urban India is undoubtedly growing at an exponential rate. As of February 2016, about 377 million people from India's total population of 1.21 billion were urban dwellers. With more than ten million people migrating to cities and towns every year, the total urban population is expected to reach about 600 million by 2031.[19] This is further compounded by the

---

[19]http://www.thehindu.com/features/homes-and-gardens/indias-challenge-of-disordered-urbanisation/article8285145.ece

concentrated number and nature of urban centres. Against this context, it is imperative that critical issues like adequate sanitation provision are addressed concurrently and the right players who understand the importance of sustainability and scale are brought in. This necessitates the involvement of the corporate sector.

Rural India has been easier to tackle with panchayats and district administrations taking the lead, often supported by NGOs and corporates. Innovations are continuously required to improve design of toilets adapting these to different local conditions. The supply chain needs to be strengthened to reach all locations.

The enabling framework to enhance corporate sector participation in sanitation would need to move beyond the conventional role of corporates envisaged in construction of toilets to tapping their expertise in behaviour change, project management, innovation and supporting disruptive technologies.

To take this beyond CSR, there is a need to develop an 'Ease of Doing Business in Sanitation' for the corporates, which can be achieved by aligning sanitation with the existing CSR strategies rather than developing altogether new ones. Examples could include creating sanitation awareness among school students under education or school programmes. With BUMT as its philosophy, the ISC works with experts who test and develop such effective communication programmes. Another way would be to develop a supporting market of specialists/aggregators/advisory companies who can hand-hold the corporates and provide end-to-end solutions. It could also include providing a business case on technology, material and services, etc. such that the value chain is inclusive and well integrated with established businesses working together to deliver a complete service of a high standard. Another way could be by supporting institutional frameworks for certifications, standards and verification, which would help to flag areas of strength, areas that need improvement and linkages between them.

Moving forward, and as is being anchored by the ISC, it is critical to first undertake a mapping exercise to understand which corporates are doing what in various capacities and where. By doing this, the learnings and experiences can be shared across the sector and a catalogue of ideas for other corporates can be created. Thereafter, corporates must be engaged for the purpose of capacity building to better understand the situation, particularly in the context of urban sanitation in India. Based on this understanding, an ecosystem can emerge to facilitate a marketplace for match-making between corporates and other stakeholders, including implementation partners. It is also important that corporates are able to align their strategic interests with that of the government's SBM.

The issue of safe and sustainable sanitation is both a health and development imperative. From a CSR perspective, it is critical to align with the country's goals and achieve the resulting outcomes of better health, education (as attendance of children, especially girls, improves) and community welfare. The current focus on sanitation, which is historically unprecedented, provides an opportunity to create strategies and business models for delivering effective, affordable and sustainable WASH solutions suitable to the Indian context. With this holistic approach, corporates become partners in achieving a Swachh Bharat.

Swachhata Doots: The Power of Volunteering

Meet Punjab's Gurpreet Singh, who believed in Mahatma Gandhi's ideology that the best way to find yourself is to lose yourself in the service of others. For many years, Gurpreet had worked, persevered and excelled at his daily job at one of HUL's factories in India. But somewhere, deep inside, he yearned to do more. He hoped to do something for his community and more importantly, his nation. That is when an opportunity came knocking on Gurpreet's door in the form of an innovative programme by his company called 'Swachhata Doot' (Messenger of Cleanliness). He swelled with pride at the thought of spreading awareness about the connection between cleanliness and eradicating diseases in his village.

The programme creatively bundled together various facets to create a strong employee volunteering programme that was both powerful and impactful. These facets were:

- A large network of twenty-nine factories and 2,000 plus factory workers located in remote locations of India.
- An innovation in media to make the message reach far and wide in media-dark areas.

- The will and desire to contribute to the nation.

The idea was to empower factory workers like Gurpreet to become messengers of cleanliness reaching out to villages with the message of Swachh Aadat. The ISC, with its vision to enable and support an ecosystem for sustainable sanitation, has approached other corporates and large organizations like the Indian Railways and with HUL's support hopes to make Swachhata Doot a national movement towards sanitation. Basically, it is a mobile-led rural BCC model. As part of this model, factory workers become agents of behaviour change in their villages by sharing two-minute audio stories on three Swachh Aadats through their mobile phones. Through its network, HUL's plan is to reach out to five million rural lives and educate them on clean habits. The ISC through its network of organizations working in the field of sanitation will be a key collaborator in this project.

WHY THIS IDEA?

The noble cause of SBM necessitates the need for a change in behaviour along with the creation of infrastructure. Repeated reports and reviews tell us that the greater challenge for India lies in changing behaviours—in getting people to actually use the infrastructure.

Every year, in India alone, 1.3 million children die before they reach the age of 5, many due to preventable infections. Over 40 per cent of these deaths occur in the neonatal period, that is, the first twenty-eight days after delivery. Something as simple as handwashing at right occasions could reduce these deaths significantly. Over the years, HUL's personal-care products have developed and refined a handwashing campaign targeting schoolchildren called 'School of 5' that has proved to change behaviour at school and the household level. Communication plays an important role in changing behaviour. However, communication strategies in villages can become tricky. With no universal access to the Internet and low TV penetration, social change becomes a challenge.

LEVERAGING THE MOBILE PHONE

More than 70 per cent of India resides in villages. Many of these villages haven't yet been exposed to media choices. Therefore, it is imperative to think about innovative means, beyond traditional media, that can take the message of SBM to the larger populace.

It is a well-known fact that India is one of the leading users of mobile phones. A recent report by the Telecom Regulatory Authority of India (TRAI) declared that the total number of mobile subscribers in India was 1.15 billion as of December 2016. Of these, 462 million were rural subscribers.

Can a mobile phone be used as a conduit for people to absorb relevant information, build conviction and possibly change mindsets? Absolutely.

HUL factory workers were trained to become messengers of cleanliness with mobile phones in their hands. The power of their belief and technology together has been able to impact attitudes on issues pertaining to open defecation and water safety.

HOW DOES IT WORK?

Each factory works out a congenial schedule that enables volunteering factory workers to set aside some time to act as Swachhata Doots. The factory workers go back to the villages they have come from (or otherwise allocated locations) and use the mobile phone to give a missed call to a number from where audio content stories on each of the three good habits of drinking clean water, using a clean toilet and washing hands with soap are disseminated. These are interesting and engaging stories intended at influencing attitudes, knowledge and practices of the communities. The programme is also designed to have multiple touchpoints in the village—the mohalla, Anganwadi and the school. This entire initiative runs smoothly, with complete ownership of the factory management team, including the Human Resources (HR) department. Right from training of the workforce on what they need to do as Swachhata Doots to executing the programme on ground is planned in a manner that there is no disruption to business. As the scale of the programme rests on existing infrastructure of factories and mobile networks—the cost of implementation is also economical. A big difference from earlier efforts is that the person communicating the message is from within the community and so acceptable, not an outsider who has always been treated with suspicion.

CREATING A MORE ENGAGED WORKFORCE

The Swachhata Doot programme has empowered the front-line workmen. As is often the case with corporate volunteering programmes, it has given them a purpose which goes beyond ensuring fruition of production plans. It fills them with a sense of pride, gives them a status—far elevated from what their 'job description' is; the recognition in the local communities instils a sense of pride that is unparalleled. At the end of the day, the factory worker feels satisfied that he did something good for the country, something that echoes with the goals of the ISC.

THE ROAD AHEAD

For the Government of India's SBM to be a success, it is important that the message of cleanliness is taken to the interior parts of India where health and hygiene matter the most. Clean habits will usher in a clean India. The innovative model of Swachhata Doot can also be used to have an impact on sustainable sanitation solutions that need to include the entire value chain of BUMT. Both the conversation and efforts around sanitation need to be viewed through ISC's lens of BUMT—and partnerships on such models are a must. Past campaigns have suffered from a lack of human resources and the capacity to implement and monitor rural sanitation uptake along with behaviour change. The daunting challenge and scale of the mission, aiming at an ODF India in the remaining time frame, requires an urgent boost in the implementation capacity.

We believe that companies have a key role in contributing to the health and hygiene agenda of the country. With HUL joining hands with the ISC, we can make this a national movement where anyone and everyone can become a Swachhata Doot. It is an invitation for everyone to participate in the movement.

The ISC with a huge base of corporates has taken this programme to them for piloting it around their factories. We need more corporates, PSUs and any organization for that matter which has human resources like the police, army, railways or even schools to sign up on such programmes. All of them can easily train Swachhata Doots—spreading the three habits of WASH. We can only imagine the magnitude of benefit this will bring to India as a national movement towards sanitation.

Corporate volunteering is a great way to build engagement of employees—and there can be no better way than to contribute to a massive national agenda. It is a win-win for the employee, the corporate and the country. Mother Teresa had rightly said: 'I alone cannot change the world, but I can cast a stone across the water to create many ripples.'

Role of NGOs and Development Agencies in Sanitation

Though the government plays a major role in the water and sanitation sector, development partners and NGOs along with corporates are critical players in the sanitation ecosystem. They are an important link between the people's needs and the government and act as catalysts of social change. To effect behaviour change and sustain it, collaboration between government and NGOs and communities is essential. They also play a key role in providing policy-level inputs to the government and design of programmes.

NGOs in sanitation have made government programmes more responsive to the needs of the people. They have helped in organizing communities to take advantage of government programmes pertaining to them as they are in closer proximity with target beneficiaries, creating trust between the government and the people. They have also played a role in empowering the communities for demanding their entitlements and capacitating them to do citizen-level monitoring.

There are largely three categories of development partners and NGOs, namely (a) Donors (b) Investors (c) Implementing Organizations. Funds for sanitation projects can come from donors like multilaterals, bilaterals, trusts, foundations, international organizations to the implementing organization. They are diverse in their roles, but in totality help through loans, technical support or grants. Today investors have started promoting social enterprises in sanitation allowing for a longer-term durability of the project. However, typically financing is provided for a limited period of time. Where long-term financial viability has not been taken into consideration, many projects may collapse once the finances are no longer available.

DONORS, MULTILATERALS AND BILATERAL ORGANIZATIONS

- **World Bank:** It is one of the key government partners focusing on the SBM. The modalities of their support have been through loans and technical assistance across various states. They are also into various capacity building programmes, institutional development, monitoring and evaluation and BCC. The World Bank loan is being disbursed in a phased manner using the Development Linked indicators (DLI). The project provides for incentivizing states on the basis of their performance in SBM Gramin (G). The total approximate amount of the loan is $1.5 billion.

  The project is expected to accelerate efforts to achieve sustained outcomes in sanitation by 2019. The incentive framework introduced through the project will reorient efforts of states towards the SBM (G) 'outcomes', such as reduction in open defecation, sustainable achievement of ODF villages and improvement in SLWM. The project has in place a robust and credible independent verification system for annual measurement of improvement in rural sanitation. The project is also supporting the SBM (G) programme in achieving its objectives of attaining an ODF and clean environment. Since poor sanitation is related to ill health, malnutrition, poor education and poverty, achievement of SBM (G) objectives will have a beneficial effect on all of these. It will therefore ensure a better quality of life for the rural population. An Expert Working Group has been established with cross-sector representation to monitor progress.

- **USAID:** It is focusing on innovation in WASH, e-learning and information and communication technology (ICT) in WASH. They are supporting the SBM (U) by first creating awareness of WASH issues and triggering demand from citizens and local government bodies for solutions in selected urban and peri-urban geographies. USAID and its partners identify scalable WASH solutions that the Government of India and the private sector can implement across the country. Some of the notable efforts of USAID are support to the Project Management Unit (PMU) along with Bill & Melinda Gates Foundation (BMGF), support to WASH alliance members like Centre for Urban and Regional Excellence (CURE), CEPT University, Urban Management Center (UMC), National Institute of Urban Affairs (NIUA), etc.
- **United Nations Children's Fund (UNICEF):** It has been a key player in the sector's drive of eliminating open defecation in India. They have developed the national

sanitation and hygiene, advocacy and communication strategy (SHACS) for the Government of India and are working with state governments to implement this. They are working with governments to establish state open defecation elimination plans and improving the efficiency of the government's rollout plans. UNICEF in India has introduced pilots of Community Approaches to Total Sanitation (CATS) in six states, to demonstrate how grassroots methods can be harnessed to deliver ODF communities quickly and with quality. In terms of mainstreaming, UNICEF is working in collaboration with the health ministry to map WASH compliance in health facilities in the most deprived districts and making recommendations to address non-compliance. UNICEF's WASH, and Advocacy and Communication sections developed the Poo2Loo campaign. This unique campaign deliberately chooses to address the population of young Indians who have a toilet at home in order to sensitize them to the plight of those who do not have toilets and create a youth social movement to stand up and advocate the need for everyone to have a toilet. The campaign, launched in the largest cities of India and in its second phase, is being spread to state capitals and smaller cities.[1]

- **GIZ:** It works very closely with the government on policy development. It has supported the development of City Sanitation Plans and School Rating Project. It is also supporting SuSanA, which is one of the biggest global platforms on knowledge sharing in sanitation, the India chapter for which is very active and hosted

---

[1] http://unicef.in/Whatwedo/11/Eliminate-Open-Defecation

by ISC with advisory members from Arghyam and Ecosan Solutions.
- **Asian Development Bank (ADB):** Based in Manila, Philippines, ADB is dedicated to reducing poverty in Asia and the Pacific through inclusive economic growth, environmentally sustainable growth and regional integration. It promotes stakeholders' participation and encourages partnerships between governments, private agencies, NGOs and communities. The ADB has recently undertaken a study on the current state of sanitation services in India and offers recommendations that can help key stakeholders to work towards universal sanitation coverage in India. Based on the study, ADB has approved a $150 million equivalent debt financing to Janalakshmi Financial Services Private Limited.
- **Japan International Cooperation Agency (JICA):** It has introduced an NGO related scheme called JICA Partnership Program which aims to encourage a joint effort of the Indian and Japanese NGOs for better services, facilities and welfare to achieve a sustainable livelihood for the communities in Asia.

FOUNDATIONS AND TRUSTS

- **BMGF:** The Foundation has invested heavily on piloting and advocating FSM and spreading the behaviour change message through its partners, which is the need of the hour. It is funding innovative technologies, like Collaborative Drug Discovery, Inc. (CDD) and Bremen Overseas Research and Development Association (BORDA), research, networking like Dasra, pilots by Water for People, EY, and exposure visits. The Gates

Foundation helped a sanitation company set up the country's first-ever community FSTP in Devanahalli, near Bengaluru, and the next one in Leh.[2] Today many state governments, like Odisha, Andhra Pradesh, Maharashtra and Tamil Nadu, have taken positive actions. The BMGF has supported MDWS in assessing rural areas in districts to rank their performance on various sanitation parameters to ensure that villages become ODF. Through its ongoing funding and technical inputs, BMGF has helped in establishing the ISC which is the national platform in sanitation for all stakeholders to partner and collaborate as well as share knowledge and document best practices. BMGF has also helped in funding and establishing the NFFSM Alliance which is the umbrella alliance for any FSM related activity across the country.

- **Tata Trusts:** They are contributing to make SBM more effective and sustainable. They focus on the sanitation value chain which includes supply chain, finance, design and usage. As part of it, they are running the 'Samajhdar' (Smart) campaign, which reinforces the adoption of healthy habits: washing hands before eating and after defecation, drinking safe water, storing water properly and always using toilets. In 2017, they provided Swachh Bharat Fellows to 600 plus districts in India in collaboration with MDWS. The fellows work closely with district collectors to implement the sanitation agenda.

INTERNATIONAL ORGANIZATIONS

- **WaterAid:** Organizations like WaterAid have been forerunners in providing innovative approaches like District-Wide approach, hygiene behaviour change, community-managed toilets and community empowerment. WaterAid commits itself to support the SBM to bring about a transformation of lives, the environment and our nation. They support this change through research, policy work and service delivery interventions through partner NGOs.
- **Water.org:** It has been acting as a bridge for providing increasing financial accessibility and has been instrumental in reaching the last mile. The organization, after proving its mettle in South India, is expanding its base in North India assisted by the ISC. ISC's partnership with Water.org has led to Punjab National Bank (PNB) agreeing to kick off water and sanitation loan products in Rajasthan and Jharkhand as part of a pan-India rollout.
- **Infrastructure Leasing & Financial Services-Education (IL&FSE):** They have designed a sanitation programme, 'Saaf & Safe' for students and community members using the ICT platform. The behavioural change programme, aided by ISC, targets schoolchildren from classes V to XII who in turn influence their parents and their communities on toilet usage, keeping water safe at source and home, personal hygiene and handwashing. The Saaf & Safe programme is a capacity building and sensitization programme which has been successfully tested in Rajasthan schools. The programme was developed

together with sanitation experts and experts in film-based teaching technology. It has also implemented the first private water supply and sanitation project in the country on a PPP basis.

- **IRC International Water and Sanitation Centre:** It is a Netherlands-based think-and-do tank that supports water sanitation and hygiene services. IRC works with national and local governments, NGOs and businesses, and local communities to realize an ambitious vision: a world where no child or adult dies of causes related to water and sanitation. IRC has been experimenting to find solutions that work by leading multi-country, multi-million dollar research programmes that tackle complex problems. For example, WASHCost, IRC's $14.5 million, five-year action research project funded by the BMGF, has gathered and shared information about the true cost of providing WASH services in Burkina Faso (Ghana), India and Mozambique. These 'life cycle costs' take into account everything from construction, finance and installation, to maintenance, repairs and eventual replacement. This helps people make informed decisions, policies and practices. They have been involved with ISC as one of the core partners on its knowledge platform called 'INSIGHTS'.

- **International Development Research Centre (IDRC):** It is a Canadian federal Crown corporation that invests in knowledge, innovation and solutions to improve lives and livelihoods in the developing world. Some of the known studies undertaken by them in the sanitation space are women's rights and access to water and sanitation in Delhi and gender and essential services in low-income communities.

- **International Water Management Institute (IWMI):** It is an international organization, which works along with partner organizations and NGOs with focus on recycling and reuse of treated waste water in urban India.

INVESTORS
- **Acumen Fund:** This is a non-profit global venture fund that uses entrepreneurial approaches to solve the problems of poverty. They invest in water and sanitation solutions ranging from drinking water kiosks to affordable toilets for slums. Their recent investment in GUARDIAN (Gramalaya Urban and Rural Development Initiatives and Network) is an example of using a creative approach to microfinance to increase access to improved sanitation in rural India.
- **Asha Impact:** It is a venture capital firm specializing in early-stage investments. The firm prefers to invest in basic infrastructure and services which include affordable housing, water, sanitation and waste management and financial inclusion. The firm typically invests with an average deployment of ₹6 crore ($0.892 million). The Asha Impact Trust conducts targeted advocacy in its focus areas and engages in initiatives that develop the impact investment ecosystem in India.
- **Villgro:** Formerly known as Rural Innovations Network, Villgro is India's oldest and foremost social enterprise incubator. Villgro funds, mentors and incubates early-stage, innovation-based social enterprises that impact the lives of India's poor. GIZ partners with Samhita and Villgro to boost corporate

engagement in the incubation of start-ups and social enterprises in multiple areas including sanitation.

IMPLEMENTING ORGANIZATIONS

- **Aga Khan Development Network (AKDN):** AKDN's Comprehensive Sanitation Initiative helps in transforming sanitation in some of India's poorest states through extensive community-focused approaches. To date, it has been able to promote positive behaviour change amongst over 2,00,000 people, and empowered approximately 60,000 households to construct toilets in Bihar, Gujarat, Madhya Pradesh and Uttar Pradesh, while also commencing work in Maharashtra and Telangana. AKDN's model for integrated block-level sanitation was recently awarded the prestigious FICCI-India Sanitation Coalition Award. Since launching its five-year Initiative in April 2015, AKDN has focused on changing existing sanitation behaviour through a range of approaches such as interpersonal communication as well as motivating communities to take collective action. By engaging directly with village leaders and opinion makers, including panchayats and other support groups, such as women's groups and school management committees, it is working to reinforce messages about the importance of sanitation and achieving an ODF environment. Moreover, to ensure communities have access to the materials and skills necessary to build good-quality toilets, AKDN is also strengthening the sanitation supply chain in rural areas. Key ways of doing this is by training masons on improved building techniques and promoting decentralized Rural Sanitary

Marts (RSMs) managed by village entrepreneurs. All of AKDN's work under the initiative reinforces the key objectives of the SBM by directly supporting the government's block- and district-level efforts to ensure sanitation access and use for all.
- **SPARC:** It is one of the larger NGOs in India focused on facilitating the creation of voice of the urban poor in the development of the city. Community toilets being run on a for-profit model by SPARC in the slums of Mumbai have demonstrated success—a model that could be replicated nationally.
- **Sulabh International:** The India-based social service organization works to promote human rights, environmental sanitation, non-conventional sources of energy, waste management and social reforms through education. Innovations include a scavenging-free two pit pour flush toilet (Sulabh Shauchalaya); safe and hygienic on-site human waste disposal technology; a new concept of maintenance and construction of pay-and-use public toilets, popularly known as Sulabh Complexes with bath, laundry and urinal facilities being used by about ten million people every day and that generates biogas and bio fertilizer produced from excreta-based plants; and low maintenance waste water treatment plants of medium capacity for institutions and industries.
- **CLTS Foundation:** It was formed by the pioneer of CLTS, Dr Kamal Kar, to meet the growing demand for a common global platform for practitioners, trainers and users of CLTS. The Foundation works closely with associates and practitioners of CLTS in different countries. It bridges the gaps between the various

development actors who have taken up CLTS in the last decade. The CLTS Foundation has played a key role in integrating the crucial component of CLTS in the overall sanitation approach across the country. It provides technical expertise at the policy level and is one of the primary capacity-building institutions in CLTS.

- Organizations like Center for Policy Research (CPR) and Research Institute for Compassionate Economics (RICE), through research and advocacy, have been giving the sector invaluable evidence-based opinions on where course correction is needed and potential problems that may arise in the future to complicate current initiatives.

As is evident from the above, not-for-profit players are crucial for the success of a nationwide and ambitious programme such as SBM. The need of the hour is a sustained collaboration between the government and non-profit organizations, allowing leveraging of each other's strengths and resources.

German Expertise in India's Rise as a Swachh Nation

The sanitation problem in India cannot be understated. While decades of sustained economic growth has made India the seventh largest economy in the world today, the provision of public services such as water, sanitation, solid waste management and drainage continue to be a challenge. According to the 2011 Census, India is home to the world's largest population of people, that is 620 million, who openly defecate.[1] With an urgent need to re-energize and remodel its approach, the Government of India launched the SBM in 2014. India can and must build partnerships with countries like Germany that have been successful in forward planning and addressing the last mile. While some collaboration between the two nations has already been undertaken, there exists vast scope for further engagement and the sharing of expertise to connect current efforts with long-term sustainability as well as gain from the experience in the management and governance of national

---

[1]Girija Shivakumar. 21 November 2013. 'Half of India's Population Still Defecates in the Open.' *The Hindu*. <http://www.thehindu.com/sci-tech/health/policy-and-issues/half-of-indias-population-still-defecates-in-the-open/article5367467.ece>

sanitation programmes.

Sanitation in the Indian context is multifaceted, layered in behavioural, social and cultural complexities. Prior to Independence, revered figures such as Mahatma Gandhi spoke about the need to improve hygiene and cleanliness in the country, notably stating in 1925 that, 'The cause of many of our diseases is the condition of our lavatories and our bad habit of disposing of excreta anywhere and everywhere.'[2] Since Independence and the initiation of planning frameworks, there have been various fronts of sanitation-focused government policy and programmes. India came on board as a signatory to the Mar del Plata Resolution of 1977 that declared the period 1981–90 as the international decade of water and sanitation, reflecting the global concern about this issue. However, the 1981 Census showed rural sanitation coverage to be a mere 1 per cent[3]. Therefore, up until the latter half of the 1990s, progress on the sanitation front in India was abysmally slow. Comprehensive efforts for achieving improvements in the rural sanitation situation began with the CRSP in 1986, a nationwide programme dedicated to rural sanitation. However, this focused purely on providing household sanitation facilities and relied mainly on subsidies and provision and did not consider 'generating demand' for household toilets; it had only a limited impact on coverage, but studies indicated low usage by households. CRSP was later reformed in 1999 to become the Total Sanitation Campaign and subsequently, Nirmal Bharat Abhiyan in 2012 that also

---

[2]Aditi Malhotra. 1 October 2015. '5 Things Mahatma Gandhi Said About Sanitation.' *The Wall Street Journal*. <http://blogs.wsj.com/briefly/2015/10/01/5-things-mahatma-gandhi-said-about-sanitation/>

[3]Government of India, Ministry of Drinking Water and Sanitation. 'Country Paper India. SACOSAN VI: 11-13 January 2016'. <http://www.mdws.gov.in/sites/default/files/india%20country%20paper.pdf>

included software components like IEC and capacity building apart from infrastructure. It was only in 2008 that the issue of urban sanitation gained attention when the Government of India came out with a National Urban Sanitation Policy. Access to improved sanitation in urban areas rose from 50 per cent in 1990 to 60 per cent in 2011.[4] With regard to rural areas, significant progress was made with a jump from a coverage rate of just 7 per cent in 1990 to 31 per cent in 2011.[5] However as is evident, a lot remains to be done.

With about half of India still defecating in the open and many households that remain unconnected to the sewage system, over 1.3 lakh tonnes of human waste is generated every day and this number is ever increasing.[6] Only 30 per cent of this waste is being treated.[7] Added to these jarring statistics are the problems of poor and insufficient systems of collection, transportation, treatment and improper disposal of solid waste. As research has continued to show, such deplorable sanitation conditions have an adverse effect on human health. It is estimated that around '37.7 million Indians are affected by water-borne diseases annually, 1.5 million children are estimated to die of diarrhoea alone, and 73 million working days are lost due to water-borne diseases each year. In 2002, unsafe water and poor sanitation contributed to 7.5 per cent of total deaths and 9.4 per cent of total disability-adjusted life years in India, according to a 2008 WHO study. Further, 45 per cent

---

[4]Government of India, Ministry of Drinking Water and Sanitation. 'Country Paper India. SACOSAN VI: 11–13 January 2016'. <http://www.mdws.gov.in/sites/default/files/india%20country%20paper.pdf>
[5]Ibid.
[6]Gates Foundation. 15 June 2016. 'Revisiting India's Sanitation Challenge.' YouTube. <https://www.youtube.com/watch?v=aAiCpd9pdMM>
[7]Ibid.

of India's children are stunted and 600,000 children under five die each year, largely because of water and poor sanitation.'[8] To fully leverage the health (reduced morbidity, mortality and improvements in nutritional status) and environmental benefits of improved sanitation, it is crucial to ensure universal access at household and institutional levels. Appropriate systems for collection, conveyance, disposal of solid and liquid waste and good hygiene practices (such as handwashing, safe use of latrines, MHM, safe handling and storage of water and food) through appropriate BCC needs to be strengthened. Thus, despite recent progress, access to improved sanitation in India continues to lag behind the MDGs target (goal 7) set for sanitation. With the renewed targets recently put forward under the SDGs, a reworking of India's national sanitation programme became imperative.

The launch of a national sanitation programme in October 2014 and the need to address the country's sanitation crisis was by no means revolutionary. However, never before has any other sanitation-focused government programme been able to sensationalize and popularize the issue to such a great extent, generate such fervour and capture attention at both national and international level. The SBM objectives are aligned with the SDGs, which urge the governments to achieve adequate and equitable sanitation and hygiene for all and end open defecation by the year 2030. The SBM strives to accelerate efforts to achieve universal sanitation coverage, improve cleanliness and eliminate open defecation in India by 2019, operating under two verticals—SBM (U) for cities and SBM (G) for rural areas. Further, the SDGs aim to improve water quality by reducing

---

[8]Population Foundation of India, USAID India. 'CSR and Sanitation in India—A Fact Sheet'.

pollution, eliminating dumping and minimizing release of hazardous chemicals and materials, halving the proportion of untreated waste water and substantially increasing recycling and safe reuse globally. Not only does SBM plan to achieve the sanitation related goals eleven years earlier, but also reflect the rising aspirations of the people and the country. The benefits of achieving SBM objectives are manifold—clean villages, towns and cities; reduction in waterborne diseases; reduced mortality arising from diarrhoeal diseases linked to poor hygiene; and higher economic growth. The World Bank estimates that 6.4 per cent of India's GDP is lost due to adverse economic impacts and costs of inadequate sanitation. Attaining SBM goals is strongly linked to achieving good health, gender equality and a cleaner environment. Already, the latest Swachhata Status Report shows an encouraging 45 per cent rural sanitation coverage in mid-2015 as against the 31 per cent coverage in the 2011 Census.[9]

It is without a doubt that the SBM has catalyzed the conversation around sanitation, right from the streets to the boardrooms of corporate India. Not only has it yielded high political will and multi-ministry involvement, it has also spurred unprecedented discourse around the issue in the country. Today, the nation is at a critical junction in the SBM programme. The government, avid for infrastructure, announced the decision to build 5.2 million toilets by September 2016, or one every second.[10] However, the danger is that the renewed focus on sanitation will also be driven purely by numbers. To move ahead,

---

[9] Government of India, Ministry of Statistics and Programme Implementation. 'Swachhta Status Report 2016'. <http://mospi.nic.in/Mospi_New/upload/Swachhta_%20Status_Report2016.pdf>

[10] 'Sanitation in India: The Final Frontier.' 19 July 2014. *The Economist*. <http://www.economist.com/news/asia/21607837-fixing-dreadful-sanitation-india-requires-not-just-building-lavatories-also-changing>

there is an urgent need to build even greater momentum around a broader understanding of what will make India truly 'Swachh'.

Germany has long served as a case in point for a nation that has managed to achieve sustainable sanitation by using forward planning to address the entire value chain of BUMT. As outlined by the German Development Cooperation (coordinated by Federal Ministry for Economic Cooperation and Development), the country pursues the following policy objectives in the sanitation sector: 'To uphold the right to adequate living conditions through adequate sanitation infrastructure; to reduce health risks from waterborne illnesses and improve the standard of health; to protect the environment, especially through the sustainable management of water resources (ground water and surface water bodies); to promote economic and social development and the development of socially and environmentally sustainable towns and cities and their surroundings.'[11] This comprehensive approach has enabled Germany to think holistically and, in turn, achieve successes such as the treatment of nearly 100 per cent of its collected urban sewage.[12] Though Germany is a water-rich country, it has repeatedly emphasized the need to ensure that all strategies and programmes set out for the sector includes targets of environmental, social and economic sustainability.

VALUABLE LESSONS

Germany is one of the largest international donors in the sector. While its initial focus fell heavily on the Middle East

---

[11]Federal Ministry for Economic Cooperation and Development. February 2009. 'German Development Cooperation in the Sanitation Sector'. <https://www.bmz.de/en/publications/archiv/type_of_publication/strategies/spezial157.pdf>
[12]Ibid.

and Africa, the country's focus has now also shifted towards other emerging countries like India. Having recently celebrated sixty years of diplomatic relations, today Germany is amongst India's most important partners for trade, investment and technology. Germany and India have already undertaken various collaborations in the water and sanitation sector, particularly in the areas of river regeneration, compact water supply for small settlements, waste water treatment systems for urban areas, energy-efficient irrigation, energy recovery from waste water and PPPs.[13] Examples of such initiatives thus far include the establishing of the Indo-German Centre for Sustainability (IGCS) that focuses on cooperation between scientists of the two countries. At the national level, Germany has initiated the project, Support to National Urban Sanitation Policy (SNUSP)-II, assisting the Government of India with various schemes that also include urban sanitation improvement like National Urban Sanitation Policy (NUSP), SBM and Atal Mission for Rejuvenation and Urban Transformation (AMRUT).[14]

Reflecting this increase in interest and engagement by Germany in India's water and sanitation sector, SuSanA recently established a formal chapter in the country with the ISC. In January 2007, the Deutsche Gesellschaft für Technische Zusammenarbeit (GTZ, German Technical Cooperation) and the Stockholm Environment Institute launched SuSanA, with an aim to improve awareness of sustainable sanitation. In India, the chapter is anchored by the ISC. The recent tie-up between ISC and SuSanA strives to create a knowledge ecosystem comprising

---

[13]'Deutsche Gesellschaft Für Technische Zusammenarbeit (GTZ). November 2008. 'Sustainable Sanitation in India, Examples from Indo-German Development Cooperation'.
[14]Ibid.

an open source library home and discussions revolving around the topic of sustainable sanitation. This collaboration coincides with the latter half of the SBM. It is an opportune time to track and report on SBM and inform the MDWS and MoUD as well as state governments (state SBM coordinators, principal secretaries, district administrators, engineers and sanitation coordinators). Through the collaboration, it is envisaged to create a dynamic interface extending beyond boundaries to share, discuss, contribute and promote sustainable sanitation systems; to be a centralized alliance for all the stakeholders of the sanitation sector; and to contribute to the achievement of current and future international development goals by promoting a systems approach to sanitation provision taking in consideration all aspects of sustainability. This partnership is a perfect example of the potential for enhanced collaboration based on sharing of expertise between the two countries in the sanitation sector.

AREAS OF SHARED LEARNING

As the IGCS, SuSanA and other examples above illustrate, Indo-German collaboration in the sanitation sector is already under way. However, given the particularly re-energized environment and conducive political, legal and institutional framework created under the SBM, the opportunity and potential for further engagement is imperative. In particular, India must learn from the expertise of the German water and sanitation sector that is centred on sustainable, resource-saving and target group-oriented approaches as well as innovation.[15] India needs

---

[15] Federal Ministry for Economic Cooperation and Development. February 2009. 'German Development Cooperation in the Sanitation Sector'. <https://www.bmz.de/en/publications/archiv/type_of_publication/strategies/spezial157.pdf>

guidance on the governance and management structures of its sanitation programmes and policy. Four possible areas of shared learning between the two countries that can be explored include the redefining of sustainable sanitation, the need for decentralized and tailored solutions, the encouraging of closed-loop approaches and the need for collaborative platforms.

Sustainable sanitation, as defined by the German Development Cooperation, includes secure, affordable and dignified access to sanitation facilities; sustainable waste water and waste management that protects people against infection and preserves the environment; and awareness of hygiene behaviour.[16] In India, recently issued guidelines by the MDWS define the criteria for declaring a village as ODF to include not just access to a toilet, but also usage of toilet and safe technology.[17] What is needed from all supporters of this national programme, therefore, is a shared understanding and commitment to provision for the required concomitant infrastructure (water, safe disposal, O&M funds) to ensure that increased demand is met with the necessary attention to all aspects around the sanitation continuum. It is imperative that India learns from the German experience of forward planning to include waste water treatment and bring to the forefront issues such as the undervaluing of O&M and sludge treatment projects whilst simultaneously building toilets.

German sanitation programmes and strategies have included a target-oriented approach, emphasizing that selected sanitation interventions must be localized taking into account the needs

---

[16]Ibid.
[17]Ministry of Drinking Water and Sanitation, Government of India. 3 September 2015. 'Guidelines for ODF Verification', Issue brief no. No.S-1101113/2015-SBM. <http://www.mdws.gov.in/sites/default/files/R_274_1441280478318.pdf>

and circumstances of the users. Recognizing that Germany has been able to uphold the quality and efficiency of their centralized waste water systems owing to its water- and capital-rich profile, developing countries must explore better suited decentralized systems. Innovative solutions and supporting entrepreneurship are critical. This notion of promoting sustainable tailored solutions to respective localities and geographies will be imperative to the success of SBM in India, particularly given the diversity of the country. Furthermore, in Germany the responsibility of water supply and waste water disposal falls on the municipalities or other public corporations. Similarly, as the SBM strives to do, India can learn about the advantages of a centralized and decentralized sanitation programme while putting the onus more on states and municipalities. Germany can also help guide India on the creation of efficient governance structures that ensure the avoidance of undesirable effects of decentralization, such as staggered or delayed funding.

Germany is well known for its promotion of a closed-loop approach in sanitation, i.e. the dealing with waste and waste water systems. In particular, its ecological sanitation (ecosan) project focuses on 'ecological sanitation systems [that] enable the recovery of nutrients from human faeces and urine to the benefit of agriculture, thus helping to preserve soil fertility, to assure food security for future generations, to minimise water pollution and to recover bioenergy.'[18] This waste-to-energy model represents the futuristic thinking of Germany's approach that will be critical to India's sanitation journey moving forward.

---

[18]Christine Werner, et al. 'Ecosan—Introduction of Closed-Loop Approaches in Wastewater Management and Sanitation—A Supra-Regional GTZ-Project', in Water in the Middle East and in North Africa, pp. 263–73. <http://link.springer.com/chapter/10.1007%2F978-3-662-10866-6_22#page-2>

Germany can provide support in sharing the knowledge and creating incentives for the use of environmentally sound, closed-loop sanitation systems. This will assist India in moving to the stage of 'sanitation plus', accruing the real benefits of ensuring universal access to safe sanitation by keeping sight of the entire sanitation value chain.

Finally, in order to achieve sustainable sanitation, collaborative platforms that promote multilevel and multi-stakeholder involvement are necessary. Germany has been particularly successful in creating PPP models in the sanitation sector wherein private companies are not permitted to provide sanitation services (including waste water treatment) directly, but instead through contracts with municipalities. For example, Gelsenwasser AG is a multi-utility privately owned public water company serving the North Rhine-Westphalia region contracted by multiple municipalities.[19] This model of PPP will be a critical learning for India's growing multi-stakeholder dialogue. In India, the government has also actively elicited the support of corporates in the programme through various channels including the setting up of the SBM Kosh as well as a Corporate Facilitation Desk to provide necessary guidance. Therefore, the SBM programme has created an avenue for engagement with the corporate sector. However, till now, discussion around private engagement in the sector has focused a lot on the numbers around infrastructure creation. Going ahead, there is a pressing need to encourage innovation across the life cycle of a sanitation programme. India must encourage more entrepreneurs in these activities including to look at profitable distributed models for waste management as done in Germany. There needs to be more attention given to financing

---

[19] <http://listofcompanies.co.in/gelsenwasser-ag/>

solutions for stakeholders, innovations should be encouraged across the value chain, actionable knowledge produced and disseminated and stakeholders supported through capacity-building initiatives. Collaborative platforms such as the ISC and further learning from the German experience will help to foster such partnerships and move towards sustainable and scalable results.

Whether with the government or other Indian players like corporates and NGOs, the Coalition can play an active role in helping German stakeholders to navigate the sector and facilitating Indo-German partnerships.

There is no doubt that India is moving in the right direction in its sanitation journey to the extent that it has been able to galvanize the country around the issue through programmes like the SBM and platforms like the ISC. It is important to recognize that India already has several strong players who have worked in the sector for many years, have the expertise in implementation and capacity building and are repositories of knowledge, and there already exist donors and corporates interested in funding. Moreover, required capital expenditure for the SBM programme is estimated to be $38,095.52 million for rural and $19,572.69 million for urban, reflecting a vast scope of opportunity.[20] What is needed with the support of partner countries like Germany is the sharing of expertise and experience in the management and governance of such vast sanitation programmes and networks. From ensuring sustainability and promoting localized tailored solutions to

---

[20]Centre for Policy Research and Confederation of Indian Industry. 'Swachh Bharat: Industry Engagement – Scope & Examples'. <http://www.cprindia.org/sites/default/files/policy-briefs/Swachh%20Bharat-Industry%20Engagement%20Report.pdf>

creating decentralized management structures, Germany can help to guide India on its mission to become ODF by 2019. To work towards total and sustainable sanitation, it is imperative to keep the focus on BUMT and in order to maximize collective contributions, to tap into the tremendous potential in creating entry points for these multiple stakeholders across the entire value chain.

Together, Germany can help India become a truly 'Swachh' nation!

Faith and Sanitation

Religion has played a key role over the years and FBOs continue to form deep-rooted local connects and act as important agents of change in translating complex discourses, such as sanitation, into understandable practices. There is room for many lines of thought, and they are all necessary to attain the aggressive targets set for the Swachh Bharat campaign.

Given the tremendous political push to construct toilets and influence change in behaviour, communities need to take up ownership, believe in the need for toilets and trust in the positive health implications of owning and using a toilet. And here is where FBOs can convince people to make lasting changes in habits and use their unique position to influence communities.

Increasingly, policymakers understand the role that temples, churches, mosques, synagogues and FBOs play in welfare reform. FBOs like Islamic Relief India, Art of Living (AOL), Global Interfaith WASH Alliance (GIWA), EcoSikh, etc. have played a role in sanitation. With their outreach capabilities, influencing power, substantial scale and presence in disadvantaged and marginalized communities, faith-based involvement has influenced behaviour change, and provided on-the-ground execution and technical support. FBOs have already demonstrated this capability in other community development projects, such as building disaster risk resilience.

LEVERAGING TRUST

With positive habits, communities can reduce stunting, deaths from diarrhoea, loss of days of employment and loss of days in school, to name a few. Construction of toilets is in full swing, but what can we do about the lakhs of people who don't want to use a toilet, even when there are amenities for a toilet available? FBOs have an important role to play in changing this and building awareness.

In a panel discussion organized during ISC's Conclave in April 2017, religious leaders from the Sikh, Muslim and Hindu faiths discussed the dire state of sanitation in the country, and provided insights into how messaging from faith leaders have altered age-old mindsets.

At the huge congregations convened, faith leaders are able to capture the imagination of large numbers of people as when they speak, people listen. With crores of Indian people still defecating in the open, and the huge cost and health repercussions to the practice, gurus such as Swami Saraswati, from the GIWA, reiterated, 'We need to come out of our houses of worship. Before you go to meditation, you need sanitation. If you don't go to the toilet, you can't focus on meditation.' Akmal Shareef, Country Head of Islamic Relief India, emphasized how the five pillars of Islam guided devotees towards cleanliness, and indicated how he advocates toilet use along with handwashing. Consistently, religious texts were used to underpin the messaging. Even though sanitation is mentioned in many religious texts, where cleanliness is correlated to godliness, many Indians continue to hold on to the notion that toilets make the home impure. Faith leaders can tap into the hearts of communities, in ways that government cannot to change these ways.

DEVELOPING A SWACHCHATA KRANTI: CASE STUDY OF 2016 SIMHASTHA KUMBH

During the 2016 Simhastha Kumbh Mela, the once-in-twelve year congregation in Ujjain, Madhya Pradesh, which was attended by over five crore Hindu devotees, WASH was a significant social cause that was taken up. Huge hoardings with photos of prominent sanyasis and sadhus were displayed under the headline of '*Swachchata Kranti: Is paavan dharti par, hum khule mein na shauch kare, garv kare aise dharti par, aur shauchalaya ka prayog kare*' (Let us not defecate in the open on this holy land, use the toilet and be proud of this land). The mass awareness campaign displayed prototypes of toilet technologies, information kiosks and exhibitions, as well as puppet shows. These puppet shows depicted Lord Shiva, saffron-clad sadhus, Maa Ganga and Maa Kshipra in attempts to get people to honour their holy rivers. These shows also worked to break misconceptions that toilets were only for women and children, and concluded with sharing of information on how the government was facilitating toilet building. During their religious discourses, Swami Avdheshanand, Swami Chidananda Saraswati and Baba Ramdev all incorporated the importance of increasing hygiene in communities and ending open defecation.

GIWA is a key partner for many development organizations, since it brings together people of different faiths together on one platform. It unites them in the struggle to provide safe and sustainable sanitation for all. During religious gatherings at Haridwar and elsewhere, GIWA is active in promoting the use of toilets vs open defecation. It also approaches the whole question of caste in a definitive manner, and in recent events has had people from the cleaning communities eating together with saints and gurus—in order to break the taboos around untouchability that these important community sanitation workers face.

In other interventions, religious leaders use religious texts and sayings to promote safe sanitation, in an effort to overcome past religious and cultural myths around household toilets and their use.

SENSITIZATION AND EMPOWERMENT: CASE STUDY OF AOL

With a presence in 155 countries in six continents, AOL, as a part of its vision to empower communities and give people a voice, has been motivating volunteers to work in water, sanitation and cleanliness issues. AOL has conducted more than 48,000 hygiene camps and 23,000 medical camps benefitting crores of people. Volunteers work closely to sensitize communities on the need for sanitation, and work to empower village heads to support the building of sanitary toilets and closed sewage systems, and access sourcing funds from the panchayat. Through targeted campaigns to build awareness in local communities, and conducting extensive capacity-building training programmes to motivate and enable community members, AOL supports the building of institutional frameworks to complement government initiatives and work.

IMPLEMENTATION: ROLE OF RAMAKRISHNA MISSION

The Government of West Bengal, supported by UNICEF, initiated the Rural Sanitation Programme in Medinipur district in 1990. The Ramakrishna Mission Lok Shiksha Parishad (RKMLSP), a reputed NGO of the state, was assigned the task of programme implementation. The NGO supports RSMs at the block level with the help of sixteen cluster-level organizations (each managing four to five blocks or RSMs) and 1,000 Youth Clubs at the village level. The Youth Clubs, according to their geographical locations, are affiliated to the respective cluster-level organization and are entrusted to propagate the programme at the grassroots. The coverage of households by sanitary toilets in the district had increased to 45 per cent by 2001 (from 4.74 per cent in 1991).

MEASURING IMPACT

Though many documented cases do exist of the work that FBOs are doing in building assets within communities, critical empirical analysis is in limited supply, with the bulk of the literature being descriptive rather than qualitative. Comparisons between different types of faith-based activities, and the factors that contribute to scale or outcome quality is also short in supply. Despite the obstacles in quantitative research, qualitative research, in the form of thoughtful explorations of how to increase effectiveness, can be extracted, though still not in an explicit form.

As of now, much of the literature relies on anecdotal accounts of participants. With so much influence, it is important to monitor the congregations, and ensure that the right sort of message is being delivered. This is a difficult task, as faith unquestionably attracts irresponsible quacks, as much as it does sensible leaders.

In the words of GIWA, 'Faith can change the world, and especially in a country like India where the majority of the population subscribes to faith.' Behaviours do not exist in a vacuum; they are a result of our beliefs and experiences. If we want to change the behaviour of crores of Indians who still defecate in the open, we need to change core values. Faith is a powerful catalyzer that plays a tremendous role in dealing with this social taboo and changing social norms.

Move to Sanitation Plus

More than three years have passed since Prime Minister Narendra Modi launched the SBM on 2 October 2014. Riding on this momentum, the nation has seen an unprecedented discourse and policy on sanitation, from vast media campaigns to the Swachh Bharat Cess on services.

The announcement of ₹11,300 crore for the SBM in the 2016–17 budget reiterated this. While the high political attention to this neglected yet critical public health endeavour is welcome, SBM has so far still seemingly focused on the number of toilets to be constructed.

There is an urgent need to build greater momentum around a broader understanding of what will make India truly Swachh. Construction of toilets will and must continue. However, we have to move forward, away from merely the provision of toilets to toilets that are used, maintained and where all human waste is safely treated and disposed of.

In this phase, which we can call 'sanitation plus'—accruing the real benefits of ensuring universal access to safe sanitation—we need to keep sight of the entire sanitation value chain to ensure sustainability of this massive national effort. As the ISC's philosophy embodies, there is a dire need to shift the focus from just build to BUMT.

Indeed, the government has clearly emphasized the need to focus on behaviour change and the usage of toilets. Guidelines by the MDWS define the criteria for declaring a village as ODF to include not just access to a toilet, but also usage of toilet and disposal of solid and liquid wastes, including child faeces. What is needed from all supporters of this national programme is, therefore, a shared understanding and commitment to provision for the required concomitant infrastructure (water, safe disposal, O&M funds) to ensure that increased demand is met with the necessary attention to all aspects around the sanitation continuum.

In order to do this effectively, all conversation and efforts around sanitation need to be viewed through a BUMT lens. Failing to do so will risk the current spends and structures built lapsing into disuse by communities that haven't been won over to consistent and universal use; leading to continued rise in diseases and deaths caused by exposure to untreated human waste in the environment. Without adequate and urgent attention to faecal sludge treatment, public health benefits that can accrue with universal access to safe sanitation will continue to elude us.

The Central Pollution Control Board (CPCB) has estimated that over 73 per cent of all faecal sludge generated in the country is left untreated in the environment in India. While the announcement of a vast ₹11,300 crore for SBM was commendable, the lower budgetary allocation of just ₹2,300 crore to the urban leg of the programme vis-à-vis the ₹9,000 crore to the rural leg continues to underplay the need for urgent attention to the issue of FSM. Unfortunately, sewage treatment is still not taken seriously in India.

It is estimated that 75–80 per cent of water pollution in India by volume is from domestic sewerage. So most of the waste that we are flushing away from our homes is making its way into water sources without treatment. The majority of the waste we excrete is going untreated into our water sources—a crisis that most of us are not aware of. Unfortunately, the rate of growth of our treatment capacity hasn't grown at the same rate and lags far behind. And even here, 40 per cent of India's total treatment capacity is located in just two cities—Delhi and Mumbai. We now have a historic opportunity to address the problem of sanitation in its entirety and use the momentum generated by SBM to realize the ambition of sustainable sanitation.

Moving ahead, we need to shift away from sporadic media coverage of toilet numbers. The CPCB statistics are alarming but conversations around sanitation continue to happen in silos. We need to bring to the forefront issues such as the undervaluing of O&M and sludge treatment projects whilst simultaneously building toilets.

We need to think long term, learning from the experiences of our neighbours like Bangladesh, to address cross-cutting issues such as how to keep our water table protected as the uptake of OSS intensifies across India. It is critical that we integrate the fragmented elements of the Indian sanitation space, both in terms of discussions and players.

There has been significant discussion around engaging corporate India through SBM. Many companies have come forward and contributed, particularly through infrastructure creation. However, in order to ensure sustained corporate engagement, we must provide an enabling framework to create a business model for sanitation that is economically viable, socially acceptable and environmentally sound. This includes harnessing the expertise of corporates across the value chain of BUMT from skilling to innovative technologies, instead of confining them to one-time contributions of building toilets. A noteworthy example is the recent move made by the government to include sanitation in the Priority Sector Lending fold. By creating an enabling framework specific to their expertise, the government has successfully onboarded financial institutions in a sustainable manner.

At the ISC, we have come across multiple players with varied strengths. Each of these players, from marketing agencies to development practitioners, is a repository of knowledge, expertise and practical insights, willing and able to engage in the Indian sanitation space. To work towards total and sustainable sanitation, we must keep the focus on BUMT. And to maximize collective contributions, we must tap into the tremendous potential in creating entry points for these multiple stakeholders across the entire value chain.

Effective Waste Management Is the Need of the Hour

Indore was recently declared the cleanest city in India. It beat 433 other cities in a survey conducted by the central government, which ranked them on various sanitation and cleanliness parameters, including waste collection, ODF status and feedback from citizens.

The survey is part of the government's initiative towards a cleaner India. Its emphasis on the issues of sanitation, open defecation and waste collection is significant, considering their impact on the environment and the health of city dwellers.

The SBM plans to achieve safe sanitation for all by the year 2019. To achieve this goal, the government has clearly defined the progress path for achieving ODF cities and districts/villages. What is more critical is a more defined process across the sanitation value chain—BUMT. Each phase comes with its suggested best practices on building, using and maintaining toilets, besides the most effective way to dispose of the waste. Needless to say, the BUMT effort needs to be sustained well after 2019.

Waste management is one of the biggest challenges that we face today. Nationally, we generate a staggering 1.75 million tonnes of faecal waste a day. However, there are no systems in place to safely dispose of the bulk of this waste. Nearly 80 per cent of this sludge—human excreta and water mixture that bears disease-carrying bacteria and pathogens—remains untreated and is dumped into drains, lakes or rivers, posing a serious threat to safe and healthy living.

A solution to this problem is the FSM system. Successfully adopted by several countries in Southeast Asia, FSM is a system that safely collects, transports and treats faecal sludge and septage from pit latrines, septic tanks or other OSS systems. This waste is then treated at septage treatment plants. FSM is an effective alternative to traditional sewerage networks—both in terms of cost of construction and the time taken. Using non-sewered sanitation systems helps to treat the bulk of waste coming from OSS such as pit latrines and septic tanks. In fact, more than 70 per cent of households with safe sanitation facilities are based on such on-site systems, and in a majority of cities there are no sewered networks or sewage treatment plants (STPs). Currently, the waste is collected by private operators of sludge-emptying services using vacu-trucks. The collected waste is however dumped indiscriminately in the nearest open space—fields, or drains, lakes or other nearby waterbodies. This poses grievous dangers of infection since the untreated sludge comes back into human contact through either the soil or the untreated water contaminated with the bacteria and pathogen load of the dumped sludge. The good news is that these truck operators can be monitored through a simple GPS tracking process in order to ensure that they dump the waste at treatment plants/pre-determined sites. Analysis has shown that treatment plants need to be conveniently located bearing in mind the need for vacu-truck operators to make money.

The FSM ecosystem requires its many stakeholders to collaborate closely. While the government will provide technical assistance to states and cities to design and implement effective FSM and treatment systems, citizens, too, need to play their part. For instance, besides paying for the services, citizens need to be aware of the importance of desludging septic tanks on a regular schedule of every two to three years depending on the capacity of the tank. They must also be ready to pay part of the cost of running faecal sludge treatment plants in their cities, through regular service charges or regular taxes.

Perhaps the most important role in the FSM chain is that of sanitation workers. From extraction and collection to transportation and disposal, they are key to an effective FSM system. At present, with no proper disposal system or safety regulations in place, they face serious health hazards. Their status in the workforce hierarchy, too, is low. However, there is huge potential in the FSM system businesses for sanitation workers. The fact is that the sludge is nutrient-rich. The waste, after treatment, can be given to farmers to be used as organic compost. It can even be treated and used for biogas, or to manufacture fuel pellets or ethanol. Once pathogens and bacteria are removed, the water can be used for irrigation, construction, and by industries in cooling plants, by Residential Welfare Societies (RWAs) and building societies for gardens and flushing and by the government for parks, etc.

With appropriate training, sanitation workers can be empowered to own and run FSM businesses, much like the producer cooperatives of the agriculture sector.

While FSM is advantageous at many levels, perhaps the most significant benefit that improved sanitation offers is public health. Cleaner waterbodies mean reduced incidence of waterborne diseases and reduced mortality linked to diarrhoeal diseases—especially among children younger than 5 years. Effective sanitation measures like FSM are critical in saving these lives.

With a national policy in place, it is now incumbent on cities and state governments to operationalize the policy. The true win of the SBM momentum will be when faecal sludge is safely managed and treated. FSM is not only an engineering or infrastructure solution, but also a city system that needs the resolve of each stakeholder to make the city faecal sludge free, and meet the objective of clean cities, as envisioned in the SBM.

A Case for Faecal Sludge Treatment Plants

Between 2006 and 2012, Bangladesh went into an overdrive to build toilets—much as India has embarked on the SBM. Its tremendous success in bringing open defecation down to under 1 per cent today (India still stands at about 35 per cent), however, has not resulted in all the health and environment benefits that were expected, primarily because in many places untreated sewage from the toilets flowed into waterbodies, increasing water pollution which causes diseases.

India is a large country with diverse patterns of human settlement, climate and natural landscape, and the above story should be cautionary—sanitation is not just about using toilets and washing hands, but requires, and includes, infrastructure to keep faecal matter and its harmful pathogens away from human contact. Toilets increase convenience and safety, particularly for women, children, the elderly and unwell. But we also need to build the appropriate and necessary infrastructure to prevent faecal matter from toilets from reaching open waterbodies and contaminating drinking water sources.

Toilets should either be connected to pipelines that take the sewage away to an STP or should be connected to properly constructed underground septic tanks that partially treat the faecal matter, a system that is referred to as on-site sanitation or OSS.

Sewerage systems and OSS require different levels of investment and effort by the government and/or individual households, and are appropriate in specific situations. Sewerage systems are difficult, disruptive and expensive to build and maintain, and only larger and wealthier cities can afford them (and even then, often only 50–70 per cent of toilets in the city can connect to these sewers). On the other hand, by 2020, over 70 per cent of urban toilets will connect to OSS systems, and these are simpler and much less expensive to build and manage.

Unfortunately, due to lack of regulations, awareness and monitoring, over 60 per cent of toilets in smaller Indian cities are either connected directly to open storm water drains and the sewage flows freely alongside our streets (breeding mosquitoes, causing a stink and overflowing onto the streets during monsoons) or toilets are connected to underground, porous 'soak pits' from which the sewage drains directly into the ground while solid faecal matter is partly retained in the pit and slowly fills it.

While toilets may provide the expected privacy, safety and comfort, these soak pits and septic tanks create two problems.

Firstly, the liquid that flows out of the tank/pit and into the ground contains pathogens, like E. coli and Hepatitis E virus (HEV), and can find their way into and pollute the underground water table or nearby lakes and rivers, causing typhoid, hepatitis, cholera and other diseases. This is already happening or is a real risk in areas where the water table is high, but because this happens underground, it gets almost no attention. Outbreaks of Hepatitis E (jaundice) in cities like Shimla (over 10,000 infected in 2016), have been traced to this cause and is drawing attention.

Second, once the septic tank or pit fills up, it chokes the toilet and has to be emptied. Cesspool vehicles, also known as 'honeysuckers' or suction trucks can be called to clean these filled-up tanks, but the incredibly toxic and dangerous 'septage' or 'faecal sludge' (both terms while technically slightly different, can be used interchangeably for the most part) that is removed from the tank or pit, is dumped into waterbodies, open land or farms—creating the obvious risks of polluted water or contaminated food supply. This is the leading cause of river and lake pollution in most parts of India.

These trucks are mostly operated by former manual scavengers, whose activities were made illegal by the Prohibition of Employment as Manual Scavengers and their Rehabilitation Act, 2013. In many rural and urban areas, however, toilets are still cleaned out manually with little or no safety gear. Skin, respiratory and other health problems abound amongst this population of sanitation workers, without whom, our cities would quickly become completely unliveable.

The first problem of liquid overflows can be addressed by building better-quality septic tanks which requires not only training contractors and masons, but also better supervision and monitoring by municipal officials who approve construction plans. Fines need to be enforced to ensure compliance. Pre-fabricated septic tanks are also available, which reduce installation time as well.

The second problem, however, requires a safe location, a treatment plant, where this septage and faecal sludge can be taken to. A few such FSTPs have been built in India in the past two years, but our 7,000 odd towns and cities need over 9,000 FSTPs to ensure that 100 per cent of our faecal sludge stays out of the environment. The total investment will be about

₹30,000 crore or $4.5 billion, and can be executed in a three- to four-year period, if there is adequate political will. Funds can be mobilized from government budgets, private investors (through PPPs) and international lenders (like ADB).

AT THE CORE OF THE SANITATION PROBLEM

Lack of effective treatment facilities sits at the core of the sanitation problem and, therefore, FSTPs are a critical component of FSM—the process of storing, transporting, treating and reusing faecal sludge and septage. This is a critical public service in any town that relies upon OSS systems, and until FSTPs are built, faecal sludge will continue to be dumped by cesspool trucks in unsafe locations.

FSM is often seen as a poorer cousin of sewerage systems, but in reality, well designed and managed FSM systems are very effective at removing the most dangerous pathogens from the environment and preventing diseases. Countries like Malaysia, Philippines and even Japan, Germany and Canada, rely on on-site solutions and FSM extensively. Major Malaysian cities have a network of treatment plants using various technologies appropriate for the volumes of faecal sludge generated locally. Through scheduled cleaning of septic tanks, the country reversed the significant water pollution in certain seaside districts that had started affecting tourism and quality of life of those who lived near the beaches

In India, this is a new area and there are currently several hurdles—government policies, technology options, financing structures, municipal priorities—that need to be understood and overcome before FSTPs can become ubiquitous. None of these are particularly difficult, but nonetheless require effort, attention and coordination of various parties including elected officials, administrators, engineers, technical advisors, banks/lenders, investors and entrepreneurs/service providers. Government functionaries, technical consulting firms or NGOs with the requisite technical skills are required to supervise such projects, so large-scale capacity-building efforts have also been started in many states.

Selecting the right treatment technology is an important decision as it determines the operational constraints, flexibility for the future and financial commitments to be made by a city. A few considerations to keep in mind:

1. **Aesthetics and odour:** An FSTP should be located inside or as close to the city as possible, so that the desludging trucks can access it quickly and easily—else, the drivers will continue to dump the faecal sludge in more accessible, nearby locations. To build any kind of waste treatment plant near urban populations, it should be clean, odour-free and hygienic, else locals complain and property prices get depressed. For example, planted drying beds are a simple and low-cost treatment solution but leave the faecal matter open to the environment and therefore are viable only in rural and distant areas. Covering them with greenhouses can address some of these concerns. More sophisticated systems are completely odourless and hygienic. In case such land parcels are just not available, FSTPs can be co-located with solid waste treatment or landfill sites. This also opens opportunity for co-composting the waste streams. The challenge can be that these sites are often far from the city and operators will find it too inconvenient to drive all the way so they will continue to dump closer to the city. Strong incentives or monitoring will be required to prevent this.
2. **Robustness:** We have seen that the quantity of faecal sludge

and its bio-chemical characteristics can vary widely not only from place to place, but even within a city depending on weather, diet and other local conditions. Monsoons often see more frequent desludging due to waterlogging, while summers can be a lean season. The FSTP should be able to deal with these variations without breaking down or compromising the efficacy of treatment.

3. **Land requirements:** Space requirements vary depending on the treatment process. Some processes, however, can be done largely underground while other plants have to be built above ground. Biological processes typically require more space but are largely underground, while electro-mechanical systems can be built on smaller plots. Land availability can thus impact choice of technologies.

4. **Waste to wealth—reuse and co-treatment:** FSTPs can produce treated water and compost which can be used for agricultural and landscaping purposes. Thermal processes including omni-processors produce biochar which is similar to charcoal and may even produce surplus electricity from the treatment process. Depending on local conditions, raw material availability and budgets, the right technology can create by-products that are required and usable locally to enhance the circular economy and recovery of wealth from waste.

5. **Cost, cost, cost:** Given that it is usually the most important practical consideration, decision-makers should consider not just initial capital cost but operating cost as well. High operating costs are the number one reason why infrastructure becomes defunct after some time—a common scenario in India. While biological systems are very inexpensive and simple to maintain, thermal systems can cost a great deal more to operate but can also produce more

valuable by-products. Each city must do a conservative risk and opportunity evaluation given its local conditions and variability, and decisions should be taken based on life cycle cost rather than initial set-up cost alone. Biological systems may cost as little as ₹1–3 crore per FSTP to build depending on the capacity, and only ₹8–25 lakh to operate each year, while thermal systems may cost upwards of ₹2 crore and annual O&M costs of ₹30–90 lakh. As technologies evolve, these costs are bound to fall.

Smaller towns may prefer lower cost, simple to manage systems, while cities which have larger and more diverse waste streams can manage and optimally run larger and more complex electro-mechanical and thermal systems. Over time, the operating cost becomes multiples of the initial construction cost; hence life cycle cost should be studied closely.

AN INTEGRATED APPROACH

FSM systems (including treatment) can be implemented for well under ₹800 per capita (₹8 crore for a city of 1 lakh people) and be up and running within three to eight months, while centralized sewerage systems cost over ₹15,000 per capita (₹150 crore for a city of 1 lakh) and can take four to six years or longer to implement. After three years of digging roads, the sewerage system in Leh is nowhere near completion, and nearly ten years after starting, the system in Puri connects less than 40 per cent of the population. In such cities, FSM and FSTPs can be rolled out quickly and economically and provide critical treatment while the full-fledged sewerage system is built and commissioned.

Today, less than 25 per cent of India's urban population

is served by sewerage systems. And even the sewerage system in a city cannot cover 100 per cent of the population due to narrow roads, topography, expansion at the edges and other reasons and, therefore, an investment in FSM is always helpful to serve this uncovered population even in the long term. Central Bengaluru has sewerage systems but needs perhaps twenty small FSTPs (each costing ₹1–2 crore) to serve fringe areas, including Electronic City.

The government could face fund and land constraints, and an integrated approach with organizations like Indian Railways, universities, PSUs and large industries, who may have funds or land to contribute, can ease bottlenecks and allow projects to move forward quickly. Steel Authority of India Limited (SAIL) in Bhilai (Chhattisgarh) and Tata in Jamshedpur (Jharkhand) are examples. In Gulbarga (Karnataka), the local agricultural college gave some land for a municipal FSTP which will also be used for research on how to promote the circular economy. PPPs can reduce the upfront funds committed by government and also increase accountability and quality of FSM services.

Pollution of waterbodies is perhaps the greatest health and environmental hazard today and as freshwater sources become scarce, it is becoming a political and security issue. Waste water treatment offers tremendous opportunity to reuse water—decentralized, small-scale STPs at the neighbourhood level or by large apartment complexes can generate treated water for 'non-contact' applications like flushing (if dual piping is provided), gardening and beautification. This treated water can also be sold to industries that need large water supply, including construction sites.

Most cities use borewells and due to excess extraction, underground water tables are dropping and becoming harder to reach. This makes it more expensive to draw water, aside from creating risks like saltwater intrusion or soil collapsing due to lack of support from the underlying water table, as is happening across Mexico City. Waste water treatment is now mandated by law, but enforcement requires the leadership of aware citizen groups and municipalities, and substantial penalties that encourage people to comply.

Citizen groups and RWAs can collect as little as ₹1,000 per person and form PPPs with the municipality and local sanitation businesses to establish FSTPs and provide timely OSS cleaning services that protect the environment and water supply.

As awareness spreads, the benefits, speed and low investment are leading cities to travel down the path towards implementing FSTPs. The entire chain of activities—licensing honeysucker operators, scheduling the cleaning of tanks and pits, enabling reuse of by-products and financing the entire operation so that it is sustainable—should be implemented as a holistic solution.

Honeysucker service providers are quite commonly found in most places and charge a fee of ₹800–1,600 for cleaning septic tanks or pits. Bengaluru alone has about 200 such trucks serving households and industries around the city. Most operators have only one to three trucks each, but a handful of larger and more professional entrepreneurs own five to seven trucks, earning profits of ₹3–4 lakh per month. It is difficult and unappreciated work. People often beat up the drivers who are suspected of polluting local areas, and the police randomly stop and fine them. Licensing and regulating players, including deploying technology like GPS tracking and sensors, will create better employment and entrepreneurial opportunities for many former manual scavengers so they can service society with greater dignity and protection.

FSM deserves greater attention as it can achieve good results with small investments. If even only 15 per cent of

India's sanitation budget (across SBM, AMRUT and other programmes) is dedicated to FSM, the date by when every Indian town will have some form of sewage management and treatment system can be brought forward by a decade or more. We can make great strides very quickly.

India's cities will grow by 50 per cent in the next fifteen years. Improved quality of urban life requires us to adopt the right policies and technologies, make the right investments and integrate entrepreneurs into the value chain more effectively.

We need to protect the environment to improve health. We need to treat waste water to address water security challenges. And we need to close the reuse cycle to build a sustainable future to house and feed our growing populations.

Making Every Drop Count

Water has defined the survival and growth of civilizations. Availability of water is and will continue to be a major driving force for meeting the growth imperatives of any nation. At the same time, for households, it would mean fulfilling the basic needs of life. India is facing a serious and persistent water resource crisis due to the simultaneous effects of agricultural growth, industrialization and urbanization, coupled with declining surface and groundwater quantity, intra and interstate water disputes and inefficiencies in water use practices. Trends indicate that by 2050, India will move into the category of a water-stressed economy. According to a recent study by Water Resources Group, water demand in India will reach 1.5 trillion cubic metres in 2030 while India's current water supply is only 740 billion cubic metres. This also implies that 40 per cent of the people in India may not have water to drink by 2030. Further, the use of water in industry is expected to grow to 13 per cent from the current 8 per cent between 2010 and 2050 as per some estimates. The situation seems worse if we consider some stark facts:

- The CPCB estimates that nearly half of the country's 445 rivers are too polluted for safe consumption and

not a single river's water can be consumed without extensive treatment.
- The UN has ranked India 120 of 122 countries for water quality, estimating that 70 per cent of the supply is contaminated. The World Bank estimates that 21 per cent of communicable diseases in India are related to unsafe water. In India, diarrhoea alone causes more than 1,600 deaths daily.
- As reported by World Resources Institute, 54 per cent of India's total area is under high to extremely high water stress and groundwater levels are declining in 54 per cent of wells across India.
- Moreover, India shares its rivers with China, Nepal, Bhutan, Pakistan and Bangladesh. Social conflicts based on water crisis have already started in the country. There is also increasing tensions between states on river linking projects.
- The impact of climate change on water availability has already been quite telling. There has been an increase in the frequency of droughts with already five droughts from 2000 till now as compared to ten drought years between 1950 and 1990. The frequency is expected to increase further between 2020 and 2049.

CHALLENGES GALORE

Most water sources are contaminated by sewage or industrial effluent discharge. A study by the CPCB says about 38,000 million litres of sewage is generated per day (in Tier 1 and Tier 2 cities), although treatment capacity exists for only about 12,000 million litres. Only 20 per cent of domestic and 60 per cent of industrial waste water is treated; the rest is discharged without treatment. Groundwater resources in vast tracts of India are contaminated with fluoride and arsenic. Fluoride problems exist in 150 districts in seventeen states in the country, with Odisha and Rajasthan being the most severely affected. The presence of arsenic in the groundwater of the Gangetic delta causes health risks to 35–70 million people in West Bengal, Bihar and Bangladesh. The increase in water pollution has resulted in an increase in waterborne diseases that leads to loss of man-days, in turn, affecting production. An Infrastructure Development Finance Company (IDFC) report states India loses ninety million days a year due to waterborne diseases with production losses and treatment costs of ₹6 billion. Although the provision of water and sanitation has improved in the past few years, it is still not satisfactory. As per the 2011 Census, around 31 per cent of rural households and 71 per cent of urban households in the country get tap water. However, this access to tap water does not ensure adequacy and equitable distribution. In India, around 18 per cent of households still have to fetch drinking water from a source located more than half a kilometre away in rural areas and 100 metres away in urban areas. The lack of municipal water supply has resulted in overexploitation of groundwater through tube wells. Resource mismanagement, underdeveloped infrastructure and unequal governance structures are at the heart of Indian water and food insecurity.

Providing safe drinking water to all in rural India is a challenging task. The MDWS has tried to meet this challenge through the National Rural Drinking Water Programme that aims to provide every rural household access to 70 litres of water per day within its premises or at a horizontal or vertical distance of not more than 50 metres by 2022. The MoUD has also implemented the Accelerated Urban Water Supply Programme for small towns in India.

A MULTIFACETED APPROACH

Efforts on water sustainability should address food, climate, agriculture, industry and ecosystems which depend on water. The Government of India is already planning interlinking of rivers for preventing floods and improving water distribution in India. However, it needs to consider the socio-economic displacements, quality of water rerouted and availability of sufficient water for interlinking. We need to promote water augmentation measures like rainwater harvesting, aquifer and groundwater recharge and expansion of water storage capacity through scientific methods. Integrated Watershed Management (IWM), efficient irrigation practices, control of water pollution, research and development with due consideration to the externalities of the project should be advocated.

For waste water management, water recycling and reuse programmes with a focus on reducing energy cost of waste water treatment should be developed. Zero Liquid Discharge (ZLD) facilities for reusing and recycling every drop of water should be encouraged based on feasibility and cost-effectiveness. Monitoring methodologies like water balance, footprinting and auditing should be developed and awareness should be created among all sectors about improved technologies being adopted by the municipality, agriculture and industry. The government should formulate compliance standards for different segments of water users.

For effective policy and regulatory framework, India needs to encourage baseline data collection for water quantity and quality and thus regulate water allocation and water pumping. We need to propose realistic pricing of water, based on investments made and introduce reward and punishments for maintaining discipline in water management. Immediate measures include policy review, ban on flood irrigation and discontinuation of free power supply to pump underground water and prevention of untreated sewage and effluent into rivers. Creation of a market for treated municipal waste water needs to be encouraged and incentivized.

An analytical framework to facilitate decision-making and investment into the sector, particularly on measures of efficiency and water productivity should be formulated to accelerate financing and investments in the water sector with novel financing models, which may involve PPP including civil society organizations and stakeholders in water resources development and conservation. Further, convergence of water development programmes may result in better outputs. The FICCI Water Mission has tasked itself with creating public discourse around such issues, policy advocacy and sensitizing industry about the best practices of its peer group on water use efficiency.

FORGING PARTNERSHIPS

However, one of the major challenges is to make people aware of the need to consume safe water. There is renewed interest in building toilets as we look at the water and sanitation programme announced by Prime Minister Narendra Modi on Independence Day in 2014. The shame of being the highest open defecating society in the world does not help our image. Worse still, the health hazards posed by this are well known as it contaminates water and adds to the risk of waterborne diseases and diarrhoea. We need to ensure that the government, NGOs and corporates work together. The government needs to support civil society organizations involved in increasing awareness and ensuring effective implementation. An integrated campaign can result in widespread information dissemination on the ways and means of preventing contamination of water sources. As an example, HSBC had partnered with three globally admired NGOs—Earthwatch, WaterAid and WWF—with a $100 million commitment to deliver water provision, protection, information and education across the world during 2012–17. All three NGOs have a significant presence in India and are implementing large-scale programmes across seven states. Earthwatch has commissioned long-term freshwater research projects at ecologically important urban waterbodies that need to be conserved in the face of increasing pressure. HSBC employees have been trained as Citizen Science Leaders (CSLs) and engaged in freshwater quality monitoring in these key urban waterbodies across the country. The Citizen Science Leaders are collecting data to help scientists with research that will inform policymaking and urban water resource planning.

WaterAid is working with partner organizations to make communities aware of their fundamental human right to safe water and sanitation. A total of 634 schools will benefit from the NGO's work under the HSBC Water Programme in India with potential propagation of the programme values from schools to communities. WWF India is doing exemplary work in Uttar Pradesh towards conservation of the river Ganga. River conservation is complex and requires a multidisciplinary, multi-stakeholder approach. WWF India's Rivers for Life, Life for Rivers initiative envisions the Ganga and Ramganga as healthy river systems rich in biodiversity, and aims to provide long-term water security to communities, businesses and nature. It has effectively brought on board government officials and citizens to make the programme a success. The city of Moradabad has seen people taking oath as friends of the rivers Ganga and Ramganga. Notably, the city's top public officials have led the programme with WWF to set an example.

We need to create many more such partnerships if we want clean and pure water for all to become a reality.

Thus solving India's water crisis will require sustained political will, strong governance structures, efficient use of public funds, effective PPP frameworks and greater awareness of stakeholders.

Role of Corporates in Water Stewardship

We often come across businesses being disrupted due to water challenges even as governments struggle to address water issues and maintain minimum standards. Short-term efforts to tackle water woes will not be very beneficial unless water is reliably delivered and sustainably managed.

Intensive efforts are being made by businesses to enhance water use efficiency and reduce their pollution potential. Nevertheless, water challenges still pose long-term viability risk. This risk can be mitigated by a shared responsibility for protecting and managing our environment and resources through a voluntary, market-based approach by businesses.

Corporate water stewardship is defined by the Alliance for Water Stewardship (AWS) as 'the use of water that is socially equitable, environmentally sustainable and economically beneficial, achieved through a stakeholder-inclusive process that involves site and catchment-based actions. Good water stewards understand their own water use, catchment context and shared risk in terms of water governance, water balance, water quality and important water-related areas; and then engage in meaningful individual and collective actions that benefit people and nature.'

DRIVING THE AGENDA

In order to drive the water stewardship agenda, a corporate should understand its own operations, water use, operating environment and the impacts of its own operations on the environment. It is important to analyse the risk holistically across operations and the supply chain. Shared risk is the basis of water stewardship and offers an opportunity to harness the shared value of water. Water stewardship starts with companies responding to their own water related risks through improvements in policies and processes and collaboration with external stakeholders, including its supply chain, to understand potential impacts and work collectively for enhancement of efficient water management including at the catchment level.

Thus, water stewardship implies that there are internal and external components to water issues and can be classified into two categories, namely within the fence activities and outside the fence activities. The initiatives by corporates for achieving compliance to regulations are within the fence activities. This may also include initiatives within the business units to go beyond compliance like process changes, technological upgradation, etc. Outside the fence initiatives are activities by businesses outside their premises for the community and the catchment as a whole. It is not only taking up responsibility to improve the situation, but also recognizing the shared responsibility for protecting and managing our environment and resources. Overall, water stewardship involves better understanding of the company's water accounting, water footprint, use efficiency, audit, demand management, risk assessment, risk mitigation, investment in watershed management, development of new tools, standards and policies and participation in local and national water policy debates.

Water stewardship initiatives provide a long-term opportunity for availability of sustainable water resources with reduced risk to businesses. It allows businesses to minimize their impact, engage and collaborate with other stakeholders to reduce collective impact and help strengthen management of resources. It provides corporates with competitive advantage by attracting highly sensitized consumers and investors. It generates a sense of responsibility among stakeholders and also creates confidence amongst stakeholders in company policies. It provides an opportunity to enhance the company's image. Overall, it allows companies to emerge as thought leaders and build competitive advantage in the space.

Many organizations have been agents of change in promoting water stewardship. The UN Global Compact's CEO Water Mandate provides clarity and guidance through reports and guidelines for collective action, accounting, terminology and public policy. The AWS offers the global benchmark in responsible water stewardship to enable companies to understand their water use within the context of a catchment and work collaboratively to achieve shared objectives. The WWF has well-established programmes to convert the business risk into collective action at the basin level. The World Business Council for Sustainable Development (WBCSD) is supporting businesses to accelerate movement for water management, from risk assessment to implementation of response strategies at the watershed level. It has also developed the Global Water Tool to identify risk and foresee potential opportunities. The FICCI Water Mission is engaged in promoting corporate water stewardship by recognizing excellence in sustainable water management every year.

INDIAN BUSINESSES STEERING WATER STEWARDSHIP GOALS

Water stewardship as a concept has started to gain pace in India. Leading businesses are making significant progress in reducing their impact on water resources through their water stewardship strategies. The India Water Stewardship Network (IWSN) has been formed to engage multiple stakeholders in better understanding and acceptance of the water stewardship framework in the Indian context and demonstrate implementable projects. The network adopts the AWS standards for defining the framework in India. Businesses are constantly striving to improve their water use efficiency through adoption of processes and new technologies and are auditing every unit of water use. They are encouraging transparency in water use by urging supply chains to also do so. Businesses in India who have taken a stride forward towards water stewardship include the Tata Group, JSW Steel, Coca-Cola, PepsiCo, Jain Irrigation, ITC limited, Mahindra & Mahindra, Nestlé, SABMiller, etc.

Major beverage companies, like PepsiCo, Coca-Cola and SABMiller, have been actively spearheading water stewardship initiatives. They have been able to reduce their water usage by 10 to 26 per cent and have improved water use efficiency by 25 per cent. They have achieved 'Positive Water Balance' by replenishing more water than extracted by the industry through rainwater harvesting, construction of check dams, pond restoration and supporting agricultural water efficiency initiatives. PepsiCo has educated farmers to adopt direct seeding for paddy cultivation in order to prevent flood irrigation, thereby reducing water use by 30 per cent. Coca-Cola has been involved with farmers for installing drip irrigation, thereby reducing water and fertilizer cost with increased yield and has helped farmers adopt laser levelling to reduce water use by 25 to 30 per cent. Moreover, they have been actively involved in providing improved water and sanitation facilities, access to safe water and community rainwater harvesting systems. SABMiller India has engaged in integrated water resources management by driving crop productivity enhancement, soil and water conservation and livelihood generation for small farm holders. It is also exploring ways to integrate bio-treated waste water from beer to improve agricultural production in India.

Industrial giants like the Tata group, ITC group and Mahindra & Mahindra have been steering the water stewardship goals. The Tata Group has taken a stride in voluntary disclosure of water usage data through its publication, *Water Footprint Assessment: Results and Learnings*[1], for twelve plants across India and their supply chains, and then assessed the sustainability of their water footprint and prioritized strategic actions for reducing associated impacts. ITC has been contributing by basin-level water security plans, mapping of local aquifers for recharge, sustainable irrigation, water cropping systems and community-based training programmes.

Further to these initiatives, businesses can provide valuable baseline data on water quality, available water quantity, groundwater depletion, etc. with expertise and technology to the government to allow informed decision-making and effective implementation of plans. They can act as agents of change by promoting best practices among stakeholders in their watersheds and also encourage the government to allocate resources to the vital water issue. Businesses have the strength to mobilize global business communities for driving corporate water stewardship initiative in their domains and bring best practices from across

---

[1] https://www.ifc.org/wps/wcm/connect/b7af9a804148ccd9b0dbb39e78015671/Tata+Industrial+Water+Footprint+Assessment-+Final.pdf?MOD=AJPERES

the world. Thus they can become enablers and facilitators of improved water management in the public interest.

OTHER STAKEHOLDERS

The initiatives for water stewardship are incomplete without the role of other stakeholders—government, financial institutions, civil society, consumers and citizens. The government should support commitment to accountability as well as develop and implement policies for promoting corporate initiatives. It should also aim at developing strategies for incentivizing good water management practices by businesses. Financial institutions can aid in driving sustainable water management goals by creating systems to assess water related initiatives by corporates and invest in water projects or fund companies that advance sustainable water goals. Civil society can help identify and implement methods for building consumers' ability to identify responsible companies. Consumers can encourage corporate water stewardship initiatives by making purchasing decisions and policy advocacy. Businesses, investors and civil society can support development of systems that allow consumers to identify and demonstrate their preference for responsible businesses. Capacity building and awareness is an indispensable exercise for all stakeholders.

New businesses first need to understand the scope and scale of water risk and then map internal operations for water footprint, water use efficiency and reduction in pollution, including supply chain operations. Based on the contextual assessment, businesses can develop a strategy to integrate water management into operations and engage with other companies, government, communities, experts and NGOs to develop collective action strategies and leverage improved performance in the value chain. Businesses can engage in water transparency systems like CDP Water Disclosure, CEO Water Mandate, AWS and other activities at the catchment level.

Water stewardship is a new concept in India, and it is important for us to encourage as we look to manage scarce water resources in the interest of the economy, communities and the environment.

Environment and Pollution

Collective Action on Climate Change

All too often, we talk about the costs of climate action—the costs of reducing emissions, the costs of renewable energy and the costs of adapting to rising seas and more extreme weather. But for a country such as India, tackling climate change promises enormous benefits—for economic growth and productivity, public health and the alleviation of poverty. It is actually a high-carbon economy that costs the most. About one in five premature deaths in India—perhaps two million each year—is caused by environmental factors. Household air pollution, from burning solid fuels, accounts for half of these.

Outdoor particulate matter pollution, from burning fossil fuels for power and transport, caused 2.5 million deaths in India in 2015, according to a report in *The Lancet* journal, and costs the equivalent of 5.5–7.5 per cent of GDP each year. Half of the world's most polluted cities are here in India, including the top four: Delhi, Patna, Gwalior and Raipur. Indeed, ahead of the Diwali festival in 2017, authorities called on the inhabitants of Delhi to refrain from the traditional fireworks, given the already alarming levels of air pollution—but to little avail.

Our energy choices are also costing us financially. Domestic energy production has not kept pace with growth in demand, which is doubling every fifteen years. As a consequence, energy imports have surged. Between 2008 and 2012, we spent an average of 6.4 per cent of GDP importing fuel from overseas. Meanwhile, poor urban planning is also holding our economy back. In less than twenty-five years, our urban population has almost doubled from 222 million in 1990 to an estimated 410 million. By 2050, it is projected to almost double again, to 800 million. Cities are the economic engine of India, accounting for two-thirds of our GDP and 90 per cent of government revenue. But uncoordinated planning, unreliable infrastructure, chronic congestion and growing travel times are constraining the economic potential of our cities.

However, many of the measures that India is taking to address climate change can help alleviate or solve these problems. In a national climate action plan, unveiled in 2015 ahead of the Paris climate talks, Prime Minister Narendra Modi pledged to reduce the amount of carbon produced for each unit of economic output and increase the share of non-fossil based power generation to 40 per cent of total capacity by 2030.

These goals make good sense. By some measures, the cost of generating power from renewable sources in India has fallen by 65 per cent over the last three years. The cost of electricity from new power plants using imported coal is projected to be 30–50 per cent higher than the cost of wind and solar power in 2030. Homegrown renewable energy will help India's balance of payments, and bring down its exposure to volatile international energy markets.

At the same time, distributed, small-scale renewable energy can deliver to rural communities the social and economic benefits of electrification much more quickly than relying solely on extending the electric grid. Solar power, potentially coupled with batteries, can improve public health, enable access to education and provide economic opportunities. The financing agenda for renewables is particularly important for India. It will also be the focus of an upcoming UNEP Inquiry.

Reducing our reliance on dirty coal-fired power plants will help improve air quality. Providing poor people with clean cook stoves would also help address indoor pollution. These benefits are local, immediate and substantial: For India, they are estimated to be worth ₹3,600 per tonne of carbon dioxide reduced. Modi is also pursuing policies to promote smart urban development. Addressing some of the problems in India's existing model of urban planning could benefit the economy and the climate. For example, compact, connected and well-coordinated cities are more energy efficient. The World Bank estimates that for every 1,000 km of new bus rapid transit lanes, 1,28,000 new jobs are created and 27,000 premature deaths from pollution and accidents are avoided, all the while simultaneously reducing greenhouse gas (GHG) emissions.

Businesses across the country are also realizing that there are clear economic wins if they take action on climate. Leading business houses like Godrej, Mahindra & Mahindra, Tata Group, Aditya Birla Group, as well as information technology (IT) powerhouses like Infosys and Wipro, are embracing evolving areas for climate action, such as science-based targets for setting ambitious emission-reduction goals, or internalizing the price on carbon. PSUs like the Indian Railways, NTPC, GAIL and Indian Oil are also leading the business charge on national low-carbon development goals.

More than forty-two of the largest businesses in the country voluntarily participate in the India GHG Program—an initiative that facilitates measurement and management of GHG emissions. The programme builds India-specific tools and spreads sectoral best practices, driving more profitable, competitive and sustainable companies. And India's leading association of business organizations, FICCI, has established a 'green bond' working group to examine how the country's debt markets can enable the financing of smart infrastructure.

Taking action on climate change offers enormous advantages to India. The government recognizes this, and its climate pledges ahead of the Paris talks were welcomed. But we can and should go further. Greater ambition on renewable energy and reducing carbon intensity would lead to greater economic benefits for India. It is in our country's interest to capitalize on the low-carbon economy. It will allow us to enjoy cleaner air in more livable cities and, hopefully, in a more stable and hospitable climate.

In the New Climate Economy

People often ask CEOs for business advice, and we would like to share an especially important message—the low-carbon economy is the biggest opportunity of our lifetimes, and businesses that fail to recognize this fact risk being left behind. In fact, when world leaders came together in December 2015 to adopt the Paris Agreement, it sent an undeniable signal to markets and businesses all over the globe about the rise of the low-carbon future. For some companies, this may have been a wake-up call. For others, it was an affirmation of a market trend they had already recognized that cleaner growth was not only possible, but also good for their bottom line.

The NCE is becoming a reality, and businesses worldwide are realizing they not only have a crucial role to play in the fight against climate change, but also have a lot to gain—$12 trillion in worldwide revenue and savings by pursuing sustainable, low-carbon business models (including billions in cost reductions or fuel savings) and a piece of the growing $5.5 trillion global market for low-carbon goods and services. Investors too are taking note, as low-carbon portfolios and green bonds are increasingly seen as smart investments. With the help of international partnerships, it is possible through ten key actions to generate better economic growth while achieving up to 96 per cent of the emissions reductions needed to avoid dangerous climate change.

TOWARDS LOW-CARBON GROWTH

Businesses have a major role to play in driving low-carbon growth. Five hundred of the largest businesses in the world alone account for 14 per cent of all global emissions. Climate commitments and actions by businesses, including in land use, renewable energy, energy efficiency and clean infrastructure are already fast gaining momentum and are booming sectors for business. In 2016, 190 of the Fortune 500 companies together saved close to $3.7 billion through their collective renewable energy and energy efficiency initiatives. It is no longer a novelty to hear of major companies like Google or Apple 'going green'—both, for instance, have publicly and proudly invested in renewable energy and energy efficiency initiatives. Both have committed to using 100 per cent renewable energy by the end of this decade with Google on track to do so by this year, and Apple by 2020. In India, clean energy delivered over $2 billion in foreign direct investment (FDI) between 2014 and 2016. It could also yield over 300,000 much-needed jobs over the next five years.

These path-breaking actions are paying off—the most sustainable companies now exceed their peers in stock market value and financial performance. The CDP Climate Leadership Index, consisting of companies taking the strongest climate action, has outperformed the Bloomberg World Index of top companies by just over 9 per cent over the past five years. And fifty-three of the Fortune 100 companies are together saving $1.1 billion annually by investing in energy efficiency, renewable energy and emissions reductions.

The profitability of doing right by climate is increasingly clear. Procter & Gamble (P&G), the major US producer of consumer goods like cleaning supplies and personal-care products, signed up to meeting science-based targets to reduce their emissions. The energy efficiency measures they have undertaken have already saved it $500 million with potential for even more savings ahead. Almost 300 other major companies have also signed on to similar science-based targets, including behemoths like Coca-Cola, Pfizer and Sony.

Cooperative initiatives can allow businesses to go even further. They can set new norms and expectations for how businesses should respond to climate issues, and can catalyze a low-carbon transformation. For example, the Tropical Forest Alliance 2020 is a PPP that aims to transform the market for key agricultural commodities. Through the alliance, companies representing 90 per cent of the global trade in palm oil (including Unilever), have already committed to deforestation-free supply chains by 2020. Effectively, we have shifted the market to make traded palm oil supply chains deforestation-free. And we are aiming to do the same with other key commodities. In another example, the Low Carbon Technology Partnerships initiative, under the World Business Council for Sustainable Development (WBCSD), International Energy Agency (IEA) and the Sustainable Development Solutions Network (SDSN), brings together about 100 companies to accelerate the development and deployment of low-carbon technologies in key fields.

Failing to act on climate change would have clear consequences for businesses. Unilever, for instance, already faces $400 million a year in increased costs due to extreme weather events and is not alone. Eighty-eight per cent of the 2,300 companies reporting to the CDP in 2014 said climate change is an operational risk. And it is only going to worsen in the years to come.

Climate risk is also important to the financial sector. The G20 has asked the Financial Stability Board to investigate the impact of climate risk and climate policy, while countries such as China are working on a comprehensive framework for a 'green financial system'. The UNEP Inquiry is working with investors, the broader business community and governments—including China and India—to create a sustainable, low-carbon financial system.

The financial sector worldwide is also steadily shifting its money to capitalize on the burgeoning market for innovative climate solutions and moving away from riskier high-carbon investments. The market for green bonds investing in renewable energy projects or other low-carbon solutions grew from $3 billion just five years ago, to $81 billion in 2016. It will likely reach $150 billion by 2017. The shift towards clean capital is also paying off for investors. Barclays, the international investment bank, found that bond portfolios with strong sustainability attributes have outperformed those with weak environmental indices over the past seven years.

TRAILBLAZING EFFORTS BACK HOME

India has an opportunity to improve the quality and quantity of its economic growth by embracing the low-carbon economy and focusing on access to water and electricity. The government's ambitious goals for renewable energy and conserving our precious water resources recognize this. A low-carbon transition will improve our productivity and allow us to enjoy cleaner air in our cities as it envisages us living in more compact, connected and coordinated cities.

Indian businesses are trailblazing in terms of how the

private sector can respond to the climate threat and reap rewards from climate-smart decisions. Over 140 leading Indian businesses—from Ambuja Cement to Jet Airways and from ITC to Godrej & Boyce—have undertaken efforts to formally measure and manage their emissions and drive more profitable, competitive and sustainable businesses through the India GHG Program. The Tata Group, one of India's largest and most prestigious business conglomerates comprising over 100 different companies and annual revenues upwards of $100 billion, embeds a life cycle sustainability analysis into all of its product development. Tata Motors, part of the Tata Group and India's largest auto manufacturer, committed to going 100 per cent renewable because it would both reduce emissions and increase financial savings. They are also planning to launch their first electric car and electric buses as well, an important step in reducing urban emissions. On this front, Tata is following the lead of the Mahindra Group, whose subsidiary automaker was India's first to bring hybrid and electric vehicles to the market in the subcontinent. Another Indian conglomerate, the Aditya Birla Group, has integrated low-carbon initiatives throughout the corporate structure that oversees all 200 workplaces, and the CEO of each subsidiary chairs a sustainability committee. This is the kind of dedicated involvement that can bring about change in wider corporate mindsets. Indian industries are moving fast and far on climate action and are reaping the benefits. Businesses all over the world, especially in developed countries, should have no excuse for lagging behind.

It is critical that climate action by companies extends beyond balance sheets into the international policy debate. In May 2015, at the Business and Climate Summit (BCS) in Paris, a group of business associations representing 6.5 million firms called for an international climate agreement, and I [Paul Polman] was proud to add Unilever's voice to the chorus. Over 1,000 major companies and investors have called on countries to implement carbon pricing.

We are seeing an unprecedented convergence around strong climate action, from government, from business and from the religious community. In May 2015, we were honoured to be part of a small group that met and briefed the Pope and several of his top cardinals on how economic growth and sustainability can go hand-in-hand. The mobilization of such diverse groups towards the same end bodes well for the future of our planet.

2015 was a critical year, the year the international community gathered to ensure a safe future by determining SDGs and our response to climate change. Smart businesses around the world increasingly recognize that their bottom line requires climate action, and they are looking to capitalize on the NCE. Are you?

The World We Are Yet to Build

India stands at the threshold of a decisive moment in its growth path. It is on the cusp of major change. As it rapidly urbanizes, improves the quantity and quality of energy for all its citizens and manages the natural resources that underpin lives and livelihoods, the possibility of sustained and sustainable growth is within its grasp.[1]

The world's most populous democracy, home to 1.2 billion people, is also one of its fastest growing economies. Its GDP is the seventh largest in the world today and has been set

---

[1] The Global Commission on the Economy and Climate is a major international initiative that examines how countries can achieve economic growth while dealing with the risks posed by climate change. The Commission and its flagship NCE project were set up in 2013. The analysis and framing which informs this article is drawn from research conducted by the NCE and its local partner institutions, including ICRIER, WRI India, Climate Policy Initiative (more about the project, partner institutions and wider research may be found at http://newclimateeconomy.net/content/about). In particular, this article draws from the 2014 country study 'India: Pathways to Sustaining Rapid Development in a New Climate Economy' (http://newclimateeconomy.report/2014/wp-content/uploads/sites/2/2015/02/NCE-case-study_india.pdf) as well as detailed studies on urbanization, energy and finance (http://newclimateeconomy.net/content/region-and-country-case-studies).

to overtake the United Kingdom by 2020, although recent estimates suggest it has already done so.[2]

But the benefits of faster GDP growth in recent years are being undermined by harmful spillovers from the current growth model. These include severe local air pollution and its damage to health, rising energy insecurity due to an increasing share of coal and other energy imports, excessive drawdown of groundwater in agriculture, and the costs of rapid but problematic urbanization, complete with peri-urban sprawl, congestion, pollution and reduced urban productivity.

Thankfully, the negative consequences of our current growth model are avoidable. If we build our cities, farms and energy systems right, we do not have to choose between continued economic growth and other priorities like health or the environment. In fact, the steps we can take to protect nature are some of the best ways to ensure that the economy keeps its strength.

The right kind of economic growth is ongoing, inclusive and sustainable. That means ensuring that our cities are those where people can breathe, move and be productive; the energy that powers industries and homes comes from cleaner, cheaper sources; and that our natural assets can continue providing the resources and environmental services on which the well-being of present and future generations depends. The policy decisions we take today can secure this more sustainable growth path. So what are they? Which sectors will they impact? And what other opportunities might prevail if we choose a smarter model of development?

---

[2]Akshay Shah. 16 December 2016. 'India's Economy Surpasses that of Great Britain'. *Forbes*. https://www.forbes.com/sites/realspin/2016/12/16/indias-economy-surpasses-that-of-great-britain/#2d6558053bc0

First, our cities. Already home to over 410 million people, India's urban population is expected to nearly double to almost 800 million people by 2050. In the next fifteen years, 75 per cent of India's national income will come from cities, and that is also where most new jobs will be created. Research has shown that better, smarter urban growth could be an economic opportunity for India worth up to 6 per cent of GDP by mid-century, with significant savings at the household level.[3] But anyone who lives in an Indian city today can already see where the challenges lie in getting us from where we are today to how we fully realize our urban potential.

For one thing, severe air pollution is a significant burden. A recent study found that poor air quality causes around 1.1 million premature deaths every year in India, overtaking China, where deaths linked to air pollution, although still numerous, have plateaued in the last few years.[4] Ten of the world's twenty most polluted cities are found in India.[5] It is disappointing that in 2017, we had still not moved the needle on poor air quality. The recent Supreme Court ruling obliging manufacturers to abide by the new vehicular emission control norms is a positive step. A push to take electric vehicles mainstream would also be welcome, as would better data about

---

[3] Meenu Tewari and Nick Godfrey. November 2016. 'Better Cities, Better Growth: India's Urban Opportunity.' New Climate Economy. http://newclimateeconomy report/workingpapers/workingpaper/indias-urban-opportunity/

[4] Geeta Anand. 14 February 2017. 'India's air pollution rivals China as the world's deadliest'. *The New York Times.* https://www.nytimes.com/2017/02/14/world/asia/indias-air-pollution-rivals-china-as-worlds-deadliest.html

[5] Mallica Joshi. 4 June 2016. 'Half of the world's most polluted cities in India, Delhi in 11th position'. *Hindustan Times.* http://www.hindustantimes.com/delhi/four-out-of-top-five-polluted-cities-are-in-india-delhi-not-among-them/story-Gn2htcLbESB3BpeYJ4mY8K.html

traffic and air quality so that solutions can be tailored for specifics.

Congestion and traffic are also hindering our urban development. For instance, traffic congestion costs in Delhi averaged almost ₹5 per kilometre for cars and ₹10 per kilometre for buses during peak periods. The capital already has over seven million cars, well over 300 cars per 1,000 people. And India is already home to the largest number of total traffic deaths of any country—1,50,785 persons were killed in 2016.

All of this adds up. If we continue our current unconnected and poorly planned patterns of urbanization, we are looking at a cost of ₹2–12 lakh crore ($330 billion to $1.8 trillion) by 2050 or even higher. This would include the increased costs of providing public infrastructure and services to dispersed urban areas, traffic congestion, air pollution, traffic casualties and health risks. Factoring in increased road and parking capital requirements would drive costs up even higher, upwards of ₹4 lakh crore ($600 billion) per annum by 2050, plus other costs such as the value of displaced agricultural farmland. And providing public infrastructure and services to more sprawled or car-dependent neighbourhoods could be as much as 30 per cent higher compared with more compact, connected locations.[6]

MAJOR POLICY ACTIONS

There are three main policy actions that could help us harness more fully the best potential of our cities and help us deliver economic, environmental and social benefits. These are:

- Reforming land regulations to manage urban expansion to improve the efficiency and effectiveness of land use. This would cover reform of regulations including

---

[6]Ibid.

overly restrictive floor space indices, maximum building heights, setback requirements, plot-coverage ratios and parking space requirements; strengthening systems for the appraisal of land values, the determination of property rights and land registration; and conducting public land acquisitions.

- Expanding sustainable urban infrastructure to encourage appropriately compact, connected and coordinated cities. This includes enhancing centrally supported urban infrastructure programmes, with a focus on multi-modal transport planning, encouraging innovation in service provision and ensuring that urban service and user fees reflect the full social costs of services provided.
- Finally, introducing reforms to strengthen urban local government, including their financing capabilities, and improving governance and accountability. This will also include clarifying the growing responsibilities of urban governments, strengthening their administrative capacity and expanding their fiscal resources.

The second area to focus on is our energy systems. Energy is at the heart of our development ambitions, not only supporting a growing twenty-first century economy but also bringing light and opportunity to the approximately 240 million people who currently lack it.[7]

Our consumption of energy—although still small per capita—almost doubled between 2000 and 2015. Looking forward, we are set to contribute more than any other country

---

[7]International Energy Agency. 2015. 'India Energy Outlook'. https://www.iea.org/publications/freepublications/publication/IndiaEnergyOutlook_WEO2015.pdf

to the projected rise in global energy demand, around one-quarter of the total.[8] Meeting this demand will come from a combination of investing in our energy supply, ramping up efforts to improve energy efficiency and, where needed, introducing pricing reform. The 2012 power outage, which lasted about two days and affected almost 700 million people, should also be a reminder that in addition to improving and updating our energy grids, we need to look at systemic governance issues across the energy sector. For each of these there are promising signs about the direction of travel but more is possible.

Our current plans for the future of energy are ambitious, but achievable. In addition to the government's goal of achieving 175 GW of total capacity by 2022, our most recent energy draft plan has forecast that 57 per cent of our total electricity capacity will come from non-fossil fuel sources by 2027—that is three years ahead of the schedule we set ourselves as part of the global climate Paris Agreement. Much of this is thanks to the plummeting prices of solar energy, down by 80 per cent in the last five years with a recent auction of solar power in India coming in at below ₹3 per kWh, nearly the lowest in the world.[9] The government's plan also appears to have found favour with major Indian companies who are responding to its signal: Adani, for instance, unveiled the world's largest solar plant in Tamil Nadu and Tata recently announced plans to generate as much as 40 per cent of its energy from renewable sources by 2025.[10]

---

[8]Ibid.
[9]Editorial. 'Solar power breaks a price barrier'. 13 February 2017. *The Hindu*. http://www.thehindu.com/opinion/editorial/Solar-power-breaks-a-price-barrier/article17292695.ece
[10]Michael Safi. 22 December 2016. 'India plans nearly 60% of electricity

AMBITIOUS BUT ACHIEVABLE PLANS

The Indian Railways—which transports about twenty-three million people, roughly the population of Australia, around the country daily and is the single largest consumer of electricity in the country—is also planning for an exciting low-carbon future. Transitioning from the current, largely fossil fuel-based energy mix to clean energy like solar and wind power could have multiple benefits ranging from reducing its own operating costs to enhancing India's overall energy security and helping achieve clean energy targets. The cost of 100 per cent decarbonization could be 26–28 per cent cheaper than a fossil fuel-based business-as-usual pathway by 2030, largely due to the expected continuing decrease of renewable energy costs.[11]

In another encouraging move, India has been phasing out subsidies to diesel and petrol. In March 2016, the government doubled its coal cess from ₹200 to ₹400 per tonne, the third time it has been increased since it was introduced.[12]

The scope for improving efficiency standards is also significant. If we develop on a low-efficiency pathway, our overall energy demand would be 40 per cent higher by 2030, a difference equivalent to our entire current usage.[13] A famous

---

capacity from non-fossil fuels by 2027'. *The Guardian*. https://www.theguardian.com/world/2016/dec/21/india-renewable-energy-paris-climate-summit-target
[11]Charith Konda, Gireesh Shrimali and Kuldeep Sharma. April 2017. 'Decarbonization of Indian Railways: Assessing Balancing Costs and Policy Risks'. Climate Policy Initiative. https://climatepolicyinitiative.org/publication/decarbonization-indian-railways-assessing-balancing-costs-policy-risks/
[12]Ed King. 29 February 2016. 'India to double coal tax under 2016-2017 budget'. Climate Home News. http://www.climatechangenews.com/2016/02/29/india-to-double-coal-tax-under-2016-17-budget/
[13]Global Commission on the Economy and Climate. October 2016. 'The Sustainable Infrastructure Imperative: Financing for Better Growth and Development'. New Climate Economy. http://newclimateeconomy.

2010 study by McKinsey found that 70–80 per cent of India's infrastructure of 2030 is yet to be built.[14] Therefore, particularly as future demand for energy services grows, incorporating efficiency measures into our upcoming infrastructure could provide a great economic and environmental win-win.

We already have several successful initiatives under our belt. For instance, we have seen a major programme to replace old, inefficient light bulbs with LEDs at the household level, as well as citywide projects to convert street lights to their more efficient LED counterparts. Over 100 cities have signed up. A recent drive to replace over 2,00,000 street lights in Delhi was billed as the world's largest such exercise.[15] Also, India's Perform, Achieve and Trade (PAT) scheme is the first of its kind in the world—a market-based approach to enhancing energy efficiency for the most intensive sectors.

But there is scope to do much more. For instance, there are major efficiency gains to be made from shifting freight transport from road to rail while also decarbonizing the railways sector. These actions could deliver substantial savings for the Indian Railways by reducing cash outflows to power rail networks and supporting infrastructure.

ADDRESSING THE REAL CHALLENGE

At the heart of securing better growth in our cities and our energy systems is the very real challenge of finance. Our

---

report/2016/wp-content/uploads/sites/4/2014/08/NCE_2016Report.pdf
[14]Shirish Sankhe, et al. April 2010. 'India's urban awakening: Building inclusive cities, sustaining economic growth'. McKinsey & Company. http://www.mckinsey.com/global-themes/urbanization/urban-awakening-in-india
[15]Press Trust of India. 9 January 2017. 'World's Largest LED Street Light Replacement Project Launched'. *India Today*. http://indiatoday.intoday.in/story/worlds-largest-led-street-light-replacement-project-launched/1/853484.html

renewable energy target, for instance, would require about $189 billion of investment, but if we were to depend only on traditional sources of finance we would face a shortfall of roughly 30 per cent.[16] This gap could be met by institutional investors, such as pension funds, insurance companies and sovereign wealth funds, but for that to happen, there needs to be greater clarity on the business propositions and investable pathways, as well as on currency and offtake risks. There is simply no real option to miss out on these kinds of investors going forward; given the scale of the investments needed and the competing demand on public resources, we will need to be smart about bolstering private investment and finance to help bridge the existing infrastructure investment gap.

Well-designed financial instruments that are a better match with private investors' requirements are a crucial piece of this. Initiatives like The India Innovation Lab for Green Finance[17] are a welcome development in this space, helping to identify, design and accelerate the pickup of financial instruments. They can include elements like a currency hedging mechanism for foreign investment, an online peer-to-peer lending platform to connect lenders with small renewable energy developers and a financing facility for rooftop solar power which can can help drive needed private investment into renewable energy.

Engaging the public sector at the right time is critical. Given the higher risks associated with earlier phases of infrastructure

---

[16] Vivek Sen, Kuldeep Sharma and Gireesh Shrimali. December 2016. 'Reaching India's Renewable Energy Targets: The Role of Institutional Investors'. Climate Policy Initiative. https://climatepolicyinitiative.org/wp-content/uploads/2016/11/Reaching-Indias-Renewable-Energy-Targets-The-Role-of-Institutional-Investors.pdf

[17] The India Innovation Lab for Green Finance. www.greenfinancelab.in

development, public finance and support can play a key role. A clear and stable policy environment and better management of the risks associated with this phase are essential. The construction phase, which is considered the riskiest, would also benefit from targeted public support. For example, loan guarantees, currency or first-loss insurance can mitigate risks and attract co-financing to get infrastructure built.

Development banks can also play a key role at the table as they bring not only concessional finance and technical expertise, but also a higher risk appetite, which comes in handy for reducing the cost of capital for newer technologies, like solar or wind that may lack a proven track record for traditional investors. India is already playing a strong role in creating the new banks—the Asian Infrastructure Investment Bank (AIIB) and the NDB—and is a significant shareholder in both ADB and World Bank. Expanding and enhancing the role of these institutions could provide significant benefits for India.

Institutions like these are especially well placed to work as a bridge between governments and private investors, and to use public finance to catalyze private financing. For instance, once an infrastructure project such as a solar plant or a public transit system reaches the operational phase, its costs and revenues are more certain and stable which reduces default risk and makes refinancing possible. At that point, ownership can shift from governments, banks and construction companies to investors with specialized expertise in operating and managing the asset. The asset should itself be securitized and sold as bonds to the private sector with the capital then ideally recycled back to finance new infrastructure investments.

Our maiden sovereign wealth fund, the National Investment and Infrastructure Fund (NIIF), could provide an excellent opportunity to boost capital market financing for infrastructure projects. And our green bond market, which started in 2015, is already the seventh largest in the world, having raised $2.7 billion. This could be encouraged further as an instrument to enhance liquidity in financial markets and unlock capital for investment, including through agreeing on common standards for them. There are also exciting financing innovations being piloted that we should exploit going forward. For example, mobile phones have made pay-as-you-go solar home systems a reality, as rural users can make their payments through their phones without ever going to the bank. An estimated five million solar home systems will be sold between 2014 and 2018.[18]

India is stepping up as a leader on the global stage on climate action and sustainable development. Today, we have a unique opportunity to prepare and build for a future that is low-carbon and climate-resilient. If we do it right, we will gain economically from liveable and productive cities and from clean sources of energy powering our homes and industries. And we will avoid incurring later costs from climate change, which the UN estimates could reach $500 billion per year globally by 2050, with potentially even higher costs later in the century.[19] From heat and sea level rise to drought and food security, India is especially vulnerable, and our poorest fellow citizens, even more so.

---

[18] The Climate Group. 2015. 'The Business Case for Off-Grid Energy in India'. https://www.theclimategroup.org/sites/default/files/archive/files/The-business-case-for-offgrid-energy-in-India.pdf

[19] United Nations. 10 May 2016. 'UNEP report: Cost of adapting to climate change could hit $500B per year by 2050'. UN Sustainable Development Goals. http://www.un.org/sustainabledevelopment/blog/2016/05/unep-report-cost-of-adapting-to-climate-change-could-hit-500b-per-year-by-2050/

But it is within our reach to get this right for ourselves and future generations. As part of our national climate pledge for the landmark Paris Agreement, we quoted Mahatma Gandhi: 'One must care about the world one will not see'. Our future—the world we are yet to build—depends on it.

Government Is Key, Business the Solution

Climate change has no borders and no bias. It does not care whether you are rich or poor, big or small. It is, therefore, being rightly called one of the largest looming global threats facing the humankind.

The beginning of June 2017 marked the pull out of the second largest emitter of GHGs, the United States, from the Paris Agreement. It now becomes even more relevant for the rest of the world to accelerate their efforts towards climate change mitigation and adaptation.

India has taken the lead by firming up its stance to follow a low-carbon growth trajectory to fulfil the aspirations of its growing economy. The ambitious climate targets of lowering the emissions' intensity of its economy by 33–35 per cent by 2030 under the Nationally Determined Contributions (NDCs) of the Paris Agreement and carrying out one of the most mammoth renewable energy expansion programmes in the world that seeks to install 175 GW of renewable energy by 2022 is reflective of the Indian government's stance—that no matter what the world does, India won't freeride on the efforts by other nations to fight climate challenge. This puts India at the centre stage of a burgeoning global superpower that seeks to only grow in harmony with nature and urges the developed world to take leadership.

CHANGING BUSINESS CLIMATE

Under the rubric of the United Nation Framework Convention on Climate Change (UNFCCC) Paris Agreement, all stakeholders, governments, businesses, citizens and civil society have embarked on the journey to chart out a strengthened global partnership to reach the targeted net zero emissions over the course of the next half century. This is reflected by the emphasis given to non-state actors in the Agreement for the first time, which calls on businesses and corporate conglomerates to partner with governments to combat climate change.

However, the governments can only contribute to a small chunk of the pie by formulating policies, giving stimulus to climate-sensitive sectors and clean energy, building and assimilating the knowledge repository of climate-friendly solutions and technologies (that innovators develop). These solutions will ultimately need to be put into action by large and small businesses, the financial world and the manufacturing industry to have a noteworthy impact to mitigate and adapt to the effects of climate change. It is in fact good news that influential global conglomerates and business houses have taken cognizance of the threat climate change poses to their survival over the longer term, and have realized the only way to sustain themselves will be to integrate climate-friendly/climate-resilient approaches of doing business.

This climate-friendly stance taken by influential companies is the silver lining. This has set the tone of these large organizations playing the role of climate evangelists to demonstrate to smaller players that an energy-efficient, clean energy, low-emissions way of doing business will not only enhance their competitiveness and lead to energy savings, but shall also augur well towards facilitating their climate responsibility and reducing their ecological footprint on the planet.

Also remarkable is the proactive approach of the private sector which is innovating business models in a way that delivers practicable and affordable climate-friendly solutions to remote areas and to people deprived of the basic needs of electricity, water and sanitation and last-mile connectivity of public service delivery. These are the real champions who have challenged the traditional approach and taken on the risks of unchartered territory. These business models have the potential to be replicated and scaled with the right policy measures by governments. The beauty of these models is that these go beyond geographical contexts and are relevant to all parts of the world where basic needs can be fulfilled through low-carbon solutions.

REACHING A NEW SUMMIT

Climate concerns have now become an integral boardroom consideration for those businesses ahead of the curve. Climate related risks will impact all sectors and require tangible actions to address these issues. A recent report demonstrated that it was vital for financial institutions to understand that addressing stranded assets and other financial risks and opportunities associated with climate change is not a one-off action but it needs to become a permanent feature of everyday decision-making.

Just in terms of basic infrastructure alone, at least $1 trillion is required every five years, half of which needs to come from the private sector. Some of the key challenges that have precluded this from happening are the mismatch between long-term assets and short-term credit provision, as well as attracting additional flows of foreign public and private capital.[1] What is needed is a clear roadmap based on strategic sector-specific needs for channelling sustainable, adequate and predictable finance across key sectors, such as waste, low-carbon infrastructure, agriculture, sustainable transportation, and to build on innovating and developing scalable and replicable climate-proofed business models to have the desired impact.

The positive news is that businesses have already embarked on this journey. However, a push will be required from governments in devising climate-friendly policies, giving stimulus to clean energy, fiscal and regulatory assistance towards developing affordable, environmentally sound technologies and finance from developed to developing nations to enable the transitions. Action must be taken now before it is too late!

This calls for a cohesive and synergized effort by all stakeholders, a mechanism where governments and businesses can come together to develop a strategy to combat climate change. One such collaborative effort is the BCS, a unique global platform that calls for climate action. The Summit, launched in Paris (2015) and held in London next (2016) brought together businesses, investors and policymakers from all over the world to mobilize the business community in support of climate action ahead of the UN climate negotiations and emphasize swifter government action on policies to help scale up climate action by businesses. The third BCS hosted by FICCI on 31 August and 1 September 2017 in New Delhi chaired by me brought this discourse to Asia, for the first

---

[1] http://unepinquiry.org/wp-content/uploads/2016/04/Delivering_a_Sustainable_Financial_System_in_India.pdf

time convening global business, policymakers, negotiators and media in the run up to UNFCCC Conference of the Parties (COP) 23.

India was able to showcase its achievements with many overseas visitors commenting that they were unaware about India's actions in this regard. The success of the government LED lighting programme in bringing down prices through government procurement and distribution to encourage use was lauded as stories to emulate, as was the LPG programme of targeting subsidies to lower-income groups and improving supply and distribution.

The discourse focused on key areas relevant not only to businesses around the globe, but was also highly contextual for the developing world. The topics ranged from urban mobility, buildings and spatial planning, clean energy, circular economy, climate finance, markets for waste, to innovative business models for mitigation and role of carbon markets. The Summit helped bind the discourse around a PPP framework where 'government holds the key while business brings the solution'.

Rivals and Partners: China and India

When Prime Minister Narendra Modi visited Beijing in 2015, he observed that China has a strong tradition of learning, citing an old saying: 'If you think in terms of a year, plant a seed; if in terms of ten years, plant trees; if in terms of 100 years, teach the people'. India too, he pointed out, shares this core belief that knowledge and learning are supreme. Leaders from both countries should look to each other as partners in collaboration and knowledge-sharing to tackle some of the most urgent challenges of our time.

While China and India may have many substantial differences, the two Asian giants also have much in common. They are the two most populous nations, as well as two of the world's largest economies. Both countries have made extraordinary strides in growth and poverty reduction. China has harnessed three decades of rapid development to lift more than 700 million citizens out of poverty. India's GDP rose by almost 9 per cent each year for nearly a decade beginning in 2003, and it has surpassed China as the world's fastest growing major economy.

These gains, however, have been slowed by high

environmental and social costs. Income inequality, for instance, poses a particular challenge: The richest 1 per cent of households in China own a third of the country's wealth, while in India, they own about 58 per cent. Spreading the benefits of growth to a wider portion of their populations will be key, which, in turn, suggests that in future, the quality of growth for both will matter more than solely the quantity. In each country, air pollution from vehicles, power plants and industry leads to more than one million premature deaths per year.

Energy-intensive manufacturing, rapid urbanization and high energy and consumption demands have made China and India the first and third highest emitters of GHGs, respectively. Recently, both have acknowledged that fossil fuels will not be able to sustain the development they need, signalling important shifts.

RISE OF AN ECOLOGICAL CIVILIZATION
In China's case, its Five Year Plan for 2016–20 indicates its intention to become an 'ecological civilization', moving away from polluting industries and towards consumption patterns that are less resource-intensive. It will be home to the world's largest emissions trading scheme as it expands seven regional pilot trading systems to the national level.

India drew on the legacy of Mahatma Gandhi to frame its climate pledge for the landmark Paris Climate Agreement. India is not only on track to achieve its renewable energy target set for Paris, it will likely do so three years ahead of schedule. While they may be at different stages in development, both countries are poised to transform their economies to deliver high-quality, resilient and inclusive economic growth. Their success will hinge on two key areas—urbanization and energy.

Urbanization drives the economy in both countries, but

Chinese and Indian cities are experiencing significant growing pains. Dangerous levels of air pollution impose a significant burden on health and GDP and have led to higher citizen awareness and action. Traffic congestion has also become a large-scale challenge—it costs Bengaluru an estimated 5 per cent of its economic output and Beijing around 10 per cent. Both countries have introduced initiatives for better urban development.

In China, thirty-six low-carbon pilot cities have set ambitious targets for carbon intensity reductions. In India, a major push on delivering better urbanization is under way through the government's 'Smart Cities' programme. Much could be learned by comparing approaches and impacts. Recent analysis of India's urbanization used satellite data of night-time lights to compare cities' urban form with their economic growth. It found that Indian cities that were more compact in 2002 experienced faster economic growth from 2002 to 2012. Perhaps a similar analysis could be done in China.

TIME TO SEIZE THE OPPORTUNITY

The G20 is a forum for the world's largest economies, including India, China and the US, to collaborate on international economic growth and finance. However, there are multiple areas where India can work with other G20 countries to accelerate the transformation to a low-carbon, resilient economy.

India should show leadership by pushing a strong agenda for fossil fuel subsidy reform at the G20. Back in 2009, the G20 agreed to phase out fossil fuel subsidies 'over the medium term', and while there has been some progress since then, it is time to set a firm deadline for a full phaseout and couple that deadline with careful and robust monitoring of progress. India is well placed to push for this since we are seeing first-hand the benefits of reforming these subsidies at home. In the last few years, India's efforts to reform subsidies for petrol, diesel, kerosene and LPG have already decreased the budget deficit, while complementary measures have protected the poor from price changes.

Taking these lessons to the global platform of the G20 will help leaders fight against the burden of global fossil fuel subsidies, which amounted to approximately $550 billion in 2014. These subsidies make bad economic sense—they drain government budgets, reduce energy efficiency and are often ineffective at reaching the poorest. By incentivizing the burning of fossil fuels, they are also a major contributor to climate change and air pollution. Simply put, the case is clear when we weigh the benefits of reform versus the costs of the current subsidies to our health, our climate and our economic well-being. India should also push the G20 to set a deadline for phaseout of these subsidies by 2025 at the latest so that the rest of the world may follow.

This green economy is already one we understand in India. Our Paris commitments, especially those on clean energy, were ambitious and we continue to drive a strong climate agenda, including in our core role with the BRICS Bank. In April 2016, the Bank launched its first four investments, worth $811 million, all for clean energy projects. And our leadership along with France under the International Solar Alliance (ISA) which was announced in Paris— a brilliant announcement and commitment by the Indian government—should be watched closely. India and China must seize the opportunity that the G20 gives to show that they are at the forefront of climate action.

SHARED LESSONS

For India and China, there is an important learning that can be shared—better, more sustainable cities offer a clear and present economic opportunity. On energy, both China and India have made great strides in advancing renewable energy and energy efficiency, but they are still largely dependent on fossil fuels. Several opportunities exist for collaboration on technology development and deployment.

China, for example, has some of the world's largest manufacturing plants for solar and wind energy, and is the leading investor in clean energy. India's renewable energy target, if met, will be almost the same amount as the world's entire installed solar power in 2014.

Both countries are in the process of reducing fossil fuel subsidies as well. China has begun an internal review and identified nine subsidies to reform. Going forward, it can learn from India's experiences of deregulating diesel and kerosene prices and operationalizing a coal cess, with some of the revenue raised going towards a clean energy fund.

The decisions these two countries make in the next few years will be enormously consequential for the planet and for global prosperity. Just imagine what they could achieve by working together.

Addressing Pollution

Air pollution takes a long time to develop, as does resolving the problem, often many years. Remedial measures require continuous and focused follow through. It requires a multi-year, year-round plan of action, and not knee-jerk and ad hoc reactions each time air quality deteriorates on a seasonal basis. The fundamental elements of poor air are with us year-round, with some seasonal variations and spikes. The root causes need to be addressed on a long-term basis.

Secondly, air pollution knows no borders. Air quality can be impacted by occurrences many miles away, from adjoining states, to countries across the seas. I was on the Al Gore Climate Reality Project recently where astronaut Rakesh Sharma talked about how as he circled earth a plume of smoke from forest fires in Myanmar was visible impacting Indonesia! So, it is useful to recognize what issues we can have a direct impact on, and which ones we cannot.

Thirdly, air pollution is created in numerous ways (power, transport, industry, dust, waste, agriculture), which therefore need multiple lines of attack on all fronts. While understanding the contributing shares of each of these sources is important, there is no point in deflecting action on any one source by getting bogged down in debating relative shares. All major sources are bad and need to be attacked.

Finally, the action plan must have quantifiable time-bound objectives against which progress can be tracked rather than a set of disparate open-ended measures. National Ambient Air Quality Standards (NAAQS) are in place, so action needs to be planned and tracked against these standards. This focus will ensure that we do not lose momentum each time we get a seasonal improvement or the rating moves from Severe to Poor.

There have been many positive measures over the past few years. Greater focus on implementation and enforcement would add further to their success:

- Introduction of BS IV fuels and a target of 2020 for BS VI fuels.
- Introduction of fuel efficiency standards for passenger cars. These now need to be introduced and enforced for trucks.
- Phasing out of petrol and diesel vehicles over ten and fifteen years old.
- Diesel price decontrol reduced the price advantage of using diesel fuel, which significantly reduced diesel car sales. Ideally there should be no price differential between the two fuels to remove any financial incentives.
- It takes ten to fifteen years for existing vehicles to be replaced so we will have older vehicles on the road for a while even if we start now—hence we need to start as of yesterday.
- There has been considerable progress in expanding the use of liquefied natural gas (LNG) and the replacement of kerosene and wood burning for cooking.
- The rise of solar mini grids in rural areas is enabling

the replacement of diesel and kerosene usage while also providing much-needed electricity.
- The use of solar water pumps is reducing the need to rely on electricity and diesel.
- The overall rise of solar and wind-powered electricity generation and continuing positive prospects are enabling a reduction in the reliance on fossil fuels.
- The ban of petroleum coke and furnace oil and a shift to cleaner fuels is necessary and must be enforced.
- Brick kilns must not be allowed to operate unless they convert to newer, cleaner and more efficient furnace technology.
- Construction pollution has been identified as a major pollutant and therefore needs to be controlled.
- Using latest building materials and technology.
- Greater attention needs to be given to urban infrastructure with timely completion of projects. For example, the two peripheral expressways around Delhi have been under construction for ten years. These must be completed as soon as possible to enable trucks to skirt Delhi and save the city from their transit emissions.
- Urban planning that encourages mass transport and mixed commercial and residential development making our cities connected and coordinated minimizing travel are important considerations.
- Introduction of Goods and Services Tax (GST) has reduced the need for trucks to be idling at toll check posts.

While each of these measures makes sense, action must be planned holistically to keep track of the overall progress being made.

The recent attempt to introduce emission standards for the thermal power industry puts in perspective the governance process for essential air quality safeguards.

Thermal power plants tend to be major sources of a range of pollutants, including particulate matter, SO_x, NO_x and mercury. This is particularly true of older plants that are not only more polluting but are also inefficient and use copious amounts of water. In December 2015, the Ministry of Environment, Forest and Climate Change (MoEFCC) introduced comprehensive emission standards for thermal power plants, and allowed twenty-four months till December 2017 for implementation. The deadline of 7 December 2017 has passed, and little has been done by the industry to comply. It must be noted that the largest owner and operator of thermal power plants are federal (NTPC) and state-owned utilities. More modern and efficient plants owned by the private sector are operating at low levels of capacity utilization, given excess capacity in the generation sector.

The recent court-driven, National Capital Region (NCR)-centric action is welcome, but this is a nationwide problem and must be tackled accordingly. While the courts have done an admirable job in announcing ad hoc measures, the executive must develop a comprehensive inter-sectoral, long-term policy with specific targets and timelines if we are to see the change that is so urgently required. Even if the Ministry of Environment and the CPCB become more visibly active, the question of their jurisdiction over activities under control of other ministries such as Power, Industry, Transport, Urban Development, Agriculture, etc. remains. Given the seriousness and complexity of the problem and the urgency of providing solutions across jurisdictions, the recent formulation of an inter-ministerial committee under the leadership of the Prime

Minister's Office (PMO) is a welcome and much needed move.

A lot remains to be done on many fronts, including:

- Introduction of stricter emission standards for all industries and enforcement by authorities as there are standards for over 100 other industries.
- Strengthening the quality of Environmental Impact Assessments to ensure proper assessment of air quality/environmental impact and the region's carrying capacity are made and appropriate control measures have been planned for.
- Increased technical and administrative capacity of the state pollution control boards and ability for these statutory bodies to operate with greater independence.
- Improved air quality forecasting capabilities and greater clarity on emergency response measures so that these kick in before episodes, rather than being introduced as an afterthought.

A key area of focus needs to be on the electricity sector and the electrification of transportation, industry and cooking. Improving grid electricity reliability will result in less reliance on diesel generators and kerosene for lighting and power. The Indian Railways have embarked on an ambitious programme to fully electrify traction, thereby eliminating diesel traction use. This, combined by the need to facilitate a modal shift of freight away from road transport, will be very beneficial.

The government has also been looking at significantly electrifying road mobility. Improving availability of electric buses will reduce the need for citizens to drive their own two- or four-wheelers. This must be backed by a rational transit-oriented development and parking pricing policy so that citizens are nudged to make positive transportation choices. A supportive manufacturing and charging infrastructure must be created over the next few years.

Industry needs to electrify itself as much as possible for heating and other process purposes and move away from furnace oil, diesel and coal. Industry also needs to maximize its reliance of solar and wind generation for its electricity needs.

The existing surplus capacity in electricity generation and the growing installation of renewable electricity should provide impetus to electric induction cookers in order to improve household air quality.

Beyond power, industry and transportation, there are three other key areas which need urgent attention and action:

- Waste
- Agriculture
- Construction and road dust

These sectors all face distinct challenges and require differentiated solutions. Appropriate citizen and business behaviour must be incentivized by providing financially viable alternative courses of action. Fines and diktats have limited effectiveness. For example, just banning farmers from burning stubble has not made a difference to farmer behaviour. What is needed are technical alternatives that make economic sense for the farmer in order to change behaviour. The raging issue around stubble burning is a case in point where solutions are available. Subsidies would be far cheaper than healthcare costs on account of pollution. Can encouraging Happy Seeders help? Farmers in Kerala have demonstrated that stubble mixed into the soil improves productivity of the next crops. Can we find solutions that incentivize farmers to do the right thing?

The municipalities' treatment of waste and disposal

need improvement. Construction and road dust as well need monitoring. Political will needs to be in place for diktats that are required, to be effectively enforced.

Industry needs to be made aware of the policy pathway in advance in order not to disrupt production. For example, the ban on petroleum coke, as desirable and necessary as it is, was enforced with no lead time provided, resulting in disruption to production and employment. That being said, where industry has been given significant lead time, as in the case of thermal power plant emission standards or conversion to BS IV, they have failed to make use of the notice provided.

Policy action must be accompanied by greater access to data and increased consumer awareness. This will be possible through installation of a far greater number of air quality monitors and appropriate dissemination of the data collected. Our role as citizens is to push accountability and also to comply and ensure compliance with rules of those around us.

Before we despair at ever deteriorating air quality in Delhi and its environs, we must remember that such situations have prevailed in other major global cities like London, Los Angeles, etc. and have been tackled successfully. Beijing too has made progress. This did not happen overnight but took several years working to a thought-out comprehensive plan. We can, and must, do so too.

Wildlife and Habitat Conservation

DIMINISHING WILDLIFE HABITATS

The biggest threat to wildlife conservation is inadequate habitat. Some people realize this, but it is not in our face like the photo of a dead rhino bleeding, with its horn chopped off. Hence, the major concern is on poaching instead of the larger threat of habitat destruction. More rhinos are born dead than those that are killed by poachers or through natural causes.

The effort on rhino conservation in Kaziranga has been phenomenal. From being practically extinct, we now have over 2,500 rhinos. Normally, one would commend such efforts; however, Kaziranga cannot support these growing numbers due to the lack of habitat.

Similarly, the limited habitat with the growing population of tigers is leading to territorial fights causing more tiger deaths and also increasing man–animal conflict as tigers are forced out of protected areas. Each tiger needs to establish its own territory, and some tigers have been known to migrate over 200 kms. Wildlife organizations are studying how corridors can be created to link habitats for tigers for them to establish new territories.

Inbreeding could also create its own problems. Overpopulation of tigers in tiger reserves, rhinos in Kaziranga and lions in Gir, is exacerbated with the addition of new births. The silting of grasslands and woodlands gradually spreading into grasslands need to be urgently addressed. Jhumming (burning grass to remove small trees and provide softer green shoots for rhinos) is now required to be supplemented with other solutions.

One answer lies in relocating rhinos to many areas where they exist or existed earlier such as Manas (Assam), Orang (Assam), Jaldapara (West Bengal) and Dudhwa (Uttar Pradesh). Interestingly, rhinos possibly existed across northern India as they are depicted in the Indus Valley civilization and Harappa drawings. There are also references to rhino skins being used for bowstrings by the Moghuls.

Experts have opined that there is a need to initially identify suitable habitats with grasslands and waterbodies, and then develop a national plan for the relocation, with an administrative set-up and manpower to secure and manage such areas effectively. Subsequently, pilots could be conducted, and based on the learnings, larger relocations could be done. As forests are a State subject, coordination would be important. Fortunately, huge tracts of pristine forests still exist, and can be converted into national parks.

It is reported that the cost of translocating just one rhino from Kaziranga to Manas, an existing rhino sanctuary, would be approximately ₹8 lakh, and the cost of relocating one tiger is approximately ₹10 lakh. Ongoing organizational, administrative and monitoring costs would be in addition. It is believed that over the next five years large numbers would need to be relocated. Attempts at relocation have been made and continue, but the numbers required are daunting, as also the funding requirements.

To popularize new parks, accommodation for tourists would be an initial need. For this, perhaps, there should be incentives for the pioneers, say for five years.

Tigers are very tough to relocate as they cause conflict in the new areas with tigers in residence and peripheral villages. Rhinos are relatively easier to relocate. However, both are in danger of being poached when left without being monitored. There is a need for all stakeholders to get together to ensure successful relocation.

Karbi Anglong, which borders Kaziranga and has more than double the area, was declared a protected area in the year 2000, when authorities realized that animals crossing the road from Kaziranga into Karbi Anglong were no longer protected. The administrative set-up for ensuring protection to animals in that area is still wanting, and animals get poached more easily there. Moreover, despite a Supreme Court order instructing the Forest Department to remove unauthorized construction inside the designated reserve areas, we see only scattered demolitions mainly on the Kaziranga side, but very few on the Karbi Anglong side. Fortunately, we have large areas like Karbi Anglong where habitats can be increased with effective administration.

Can roads and rail tracks be realigned to skirt the periphery of parks rather than cutting through them? Where this is not possible, both overpasses and underpasses for animals could be constructed to interconnect the two sides. Internationally, it has been seen through camera traps that over a period of time, animals use these passes thus increasing their habitat, and preventing accidents.

The Wildlife Protection Act of 1972 established sanctuaries, national parks and tiger reserves numbering over 660 till 2015. This Act and Project Tiger have not only benefitted tigers, but also conserved the biodiversity of the forests and all species residing in those parks. By creating national reserve forests, we have also protected waterbodies and rivers that form part of these areas.

Ecological processes take centuries to evolve. Numerous 'highlands' are being planned to provide animals safety above flood levels in Kaziranga. Though highlands sound logical, some experts question the requirement, as some animals die in the floods but the majority survive because of well-honed natural instincts. There is also fear that on highlands animals may remain marooned and may not be able to survive till the water levels recede.

But interventions require proper studies and adequate time for evaluation through pilots, as there could be unintended consequences, e.g. sometimes flyovers on highways have changed discharge routes for water, often resulting in flooding.

An example of initiatives without proper studies and research having unintended negative results, was the offer by some parks of a bounty of ₹50 for every wild dog killed till the early '80s. The idea was to preserve prey like spotted deer for tigers. Little realizing that by killing wild dogs, the spotted deer from the open spaces were no longer being forced back by them into the forest, where tigers found it easier to prey on them. Since the bounty was renounced, it has been noticed that the spotted deer are now much more evenly spread through the forest.

Most parks have an uphill task of addressing threats from weeds which are also affecting habitats—wild rose and water hyacinth in Kaziranga, hedge blossom (or besharam, as it is locally known in Satpura, Madhya Pradesh) and lantana in several parks.

The good news is that solutions exist to increase habitat and conserve our wildlife.

GUARDIANS OF OUR FORESTS

It is heartening to note the increasing interest of people visiting game parks. As safaris become popular family holidays exposing children to forests and wildlife, it bodes well for the future. Our initial search for the elusive tiger led us to appreciate all that the forest has to offer in addition to the tiger. Today birdwatching groups are more evident and there is a growing interest in the variety of flora and fauna. However, when we returned from a game drive in Satpura some tourists asked our guide: '*Kya dekha?*' (What did you spot?) He told them that we had a fantastic sighting of the '*do bacche wali ma*' (mother with two babies)—a bear with her three-month-old twins playing on her back. We heard the tourist inform the rest of the group that he only managed to sight a bear and not tigers, not realizing that bear sightings (and with two babies) are as rare if not rarer than tiger sightings!

This was the bear's third litter and she was quite comfortable crossing the road through a number of vehicles. This augurs well for bear sightings in future, as the cubs will also be comfortable with vehicles and human movement. The tigress, Queen of Ranthambore, Machli, was very comfortable with vehicles and taking her cubs through the cars, which showed the babies that vehicles were not a threat. A large number of following generations are also comfortable with vehicles, something most animals are shy of.

With appropriate training of naturalists, guides and drivers, we can improve the tourists' overall experience beyond just chasing the tiger, encouraging interest in the varied bird life, fauna and foliage our forests offer. It is encouraging to see our guides and drivers pick up litter, supporting a litter-free environment and also ensuring visitors obey the rules. Internationally, naturalists, guides and drivers need to go through tough examinations before being allowed into parks. Has the time come for us to seriously consider the same? Guides could be made responsible for tourists in their vehicle adhering to jungle etiquette and the dos and don'ts, like talking softly, no littering, etc. For drivers, there needs to be a special test as they drive on difficult terrain and safety of tourists needs to be ensured. Also, there needs to be training to ensure respect for other tourists' sighting and photography even while showing the animal or bird to tourists in their vehicle.

We have heard of an internal conflict within the forest service, where the State Forest Service personnel are at times treated like adopted children by the Indian Forest Service (IFS). Forest guards and Range Officers spend most of their careers on their beat in the forest. Given their grassroots experience and knowledge, it would be a shame not to include their inputs in the decision-making process.

The Forest Department has done good work in terms of park management—addressing poaching; handling man–animal conflict including relocating villages out of the park; providing water holes in summer; compensating for livestock killed; maintaining roads, fire lines, fighting weeds; and regulating tourist routes, entries and exits. However, a lot more needs to be done in terms of determining problems that are faced by the different stakeholders and to be more in touch with communities around the park. Unfortunately, forest officials are often distracted by catering to VIPs, who often consider themselves above the rules.

Talking to different naturalists, guides, drivers and resort owners, we were surprised to learn that the Forest Department was not tourist-friendly. So is there a need to educate the entire chain of forest officials as to the benefits of tourism? Effective Field Directors and Divisional Forest Officers

(DFOs) periodically meet sarpanches, resort owners, guides and drivers to receive independent views on the challenges faced and their suggestions, rather than just getting the Forest Department's view. The community appreciates their grievances and suggestions being considered.

We have often tried to put ourselves in a farmer's shoes whose livestock is killed by large cats, or whose crops are destroyed by wild boar, elephants, deer, blue bull, etc. putting an end to numerous dreams and aspirations for the family. So would we react like most people and tell ourselves that poisoning these animals is not right? Would we not protect our assets? This is where the government and the Forest Department need to ensure that monetary losses are well compensated and in a timely manner. In South India, some parks have a system of recording the loss on mobiles and facilitating the follow-up paperwork that would be required. In the Anamalai Hills, they have an early warning system to alert communities of elephant presence.

The Forest Department could consider extending routes to include buffer zones and also promote regulated night drives there. This would be a novelty, and would provide free patrolling and revenue to the park. There would also be additional earning opportunities for surrounding communities, especially as most parks are open for just six or seven months in the year.

The equipping of our forest guards and attracting new recruits is in dire need of attention. The job of a forest guard, despite being a government job (normally sought after), is not a career of choice. Our forest guards, who live away from their families under very difficult circumstances, are not well-looked-after. Often, they don't have protective gear, like boots, raincoats and warm jackets for winter, and yet regularly and diligently patrol their beat where they are also injured by animals and sometimes killed by them, apart from poachers. NGOs are coming forward to provide needs like solar lights, cycles and insurance apart from fulfilling the shortfalls. Fortunately, walkie-talkies have been introduced, which helps in medical or other emergencies.

Recently, a BBC video became quite controversial as it insinuated that forest guards had a licence to kill, which created a huge backlash. It is important to note that an enquiry is held every time a guard uses live ammunition. Also, 'Halt who goes there?' will not work and the poachers will probably shoot the forest guard. Who needs to be in the park after dark unless they are up to no good? All those who live on the periphery of the park are well aware of the rules. The guards are armed with .303 rifles of World War II vintage and take on poachers with weapons like AK-47 rifles and the latest communication equipment.

Poaching and trade in animal parts are known to fund terrorists and anti-national elements. A poacher, like any stealth operator, needs to know when an animal is in a poachable area. It is not as if poachers decide one evening to pick up a gun and go on a shoot. The logistics of shooting/killing an animal, removing the body parts and getting out without detection is a complicated exercise with precise planning and needs intelligence and insider support from villages. Drones and telephone intercepts, amongst other methods, can be used to convert the hunters to become the hunted. Efforts of the Forest Department must be integrated with the police and paramilitary forces, who are the experts in anti-terrorist action.

Our forests and wildlife are only as secure as the administration that protects it. Through partnership and collaboration amongst all stakeholders, we could support and strengthen the effort.

THE CASE FOR A WILDLIFE COALITION

There are a number of organizations and parks doing excellent work in wildlife. Hence, it becomes important to establish a wildlife coalition to improve the impact of individual efforts by providing a platform for all stakeholders to come together to share, learn, collaborate and partner. This platform could also provide a convening opportunity for experts and researchers to meet and share their learnings. A wildlife coalition becomes a meeting point, both virtual and physical, for all players in the field as indeed we have seen at the ISC.

A number of organizations are spending reasonable amounts on different wildlife related activities, including helping forest guards. The coalition could help prioritize the requirements of different parks, and guide the disbursement of available funds. Raising funds from different sources, including CSR, would be another function of the coalition. A coalition of stakeholders—including government, NGOs, individual experts, businesses that service the industry, media, corporates and donors—will add greatly to the overall wildlife effort. They could facilitate forest departments, NGOs, resort owners, naturalists, guides, drivers, community and farmer representatives, sarpanches, district magistrates, etc. working together at the local level.

In such a coalition, a steering committee would guide the work through task forces, which would be established to address the different requirements of wildlife, e.g. habitat, relocation, poaching, park administration/forest guards, etc. The first hurdle would be to raise funds for a secretariat that would support the coalition. Most coalition members would have regular jobs, and would not be able to devote a large part of their time to the coalition, so the secretariat would play a crucial role in ensuring continuity of the various activities. The coalition can sensitize media to the realities of wildlife and its protection and can attempt to involve them in advocacy for conservation. It could encourage experts amongst its members to write columns in different publications and join panel discussions on TV to disseminate knowledge on wildlife.

Helping local communities understand the advantages of tourism is key to ensuring that a park does well, making them party to the conservation effort thus helping attract more tourists. More employment opportunities would arise such as of guides and drivers, and provision of vehicles and their maintenance, employment in lodges and local farmers providing food and vegetables. Villagers can also benefit from tourism as in Brazil and Africa with exposure to their villages, homes, culture and sale of their arts and crafts.

Forest related work would also provide jobs to local communities through creating fire lines, water holes, nurseries, check dams and maintaining of roads, etc. Tourism can also support the Forest Department's effort in park management through better policing by providing many more eyes and ears and feedback. The coalition can look at ways to encourage tourism with minimal negative impact on the environment.

It is surprising that often communities around the parks have not been into the park. Park visits and education on conservation for community groups, including children, would help their appreciation of the natural heritage. The coalition could help develop curriculum through experts to educate children in interactive and fun ways.

Wildlife enthusiasts communicate through informal channels and sharing of pictures. It may be worth encouraging the process and have tourists and guides post their pictures on the coalition's website with dates. Each animal can then be separated and its behaviour pattern mapped, helping the Forest Department. The coalition's website can also disseminate best practices, conversations around conservation, skilling and administration for others to emulate. Pictures would also be an advertisement for wildlife and the parks.

The coalition could arrange stakeholder workshops for capacity building, and help to identify and address issues in different parks. Information sharing and learnings from others' experience could cover wide-ranging subjects, including poaching, habitat, park administration, buffer and night drives, compensations and medical facilities. The workshops would also enable the experience and knowledge of forest guards/range officers, who work on the ground, to be captured, shared and documented.

Currently, there are a number of awards that recognize different aspects of conservation on the environment and wildlife. Perhaps the coalition should attempt to recognize the winners and support and showcase their efforts in an ongoing manner. It could also suggest to the different organizations to include awards in areas which may not have been covered or institute such awards.

The coalition can identify appropriate technologies for implementation, e.g. camera traps with flash are still being used—these scare animals, forcing them to change routes skirting the traps. Best practices in such situations can be agreed and implemented.

Project Tiger fortunately helped to protect our pristine forests, and other animals, birds and reptiles benefitted in the process. Gibbons, pangolins and other endangered species also need to be protected. The coalition can also help drive this effort.

Collaboration and partnerships are the need of the hour and a coalition could greatly enhance the overall impact of individual efforts.

Green Economy
and
Finance

The Future Path for Green Finance

The 2015 United Nations Climate Change Conference (UNCCC) COP 21, held in Paris, resulted in an agreement among the parties to limit global temperature increase to 2°C, and strives for a limit of 1.5°C if possible. Following the agreement, India set some ambitious targets to meet these climate goals. To enhance the contribution of non-fossil fuels in total power capacity and thereby reduce carbon intensity of electricity generation in the country, the renewable energy capacity was targeted to be increased to 175 GW by 2022. This would include 100 GW of solar power, 60 GW of wind energy and 15 GW of biomass and hydro energy.

As of 31 March 2017, India has achieved a third of this target at 58 GW, with wind power leading the way, and huge strides being made in solar power via auctions for solar parks. In order to achieve the 2022 targets, India will have to add about 23 GW of renewable energy capacity every year for the coming five years. If the momentum can be maintained, India would be able to achieve a renewable energy capacity of at least 260 GW by the end of the coming decade.

An addition to renewable power capacity of such massive proportion would undoubtedly need unprecedented financial investment. Both public and private capital is required to advance setting up clean energy infrastructure in the country. It has been estimated by various agencies that between 2016 and 2022, a total investment of close to $200 billion will be required, which includes $60 billion in equity and $140 billion in debt.

India faces the triple imperative of meeting its growing energy needs, extending access and improving environmental performance of its power sector. The Government of India is playing an active role on this front. To promote adoption of renewable energy resources and development of the sector, it has been offering various incentives, such as generation-based incentives, capital and interest subsidies, viability gap funding, concessional finance and fiscal incentives. The recently published report of EY has placed India at the second spot on the list of world's most attractive renewable energy markets after China.

In the three years between 2013–14 and 2015–16, India has attracted $14 billion for renewable energy investments out of which wind power projects received $7 billion and solar power projects attracted $4.5 billion. The government made capital investments and gave incentives to the tune of $1 billion. Other than this, the World Bank has extended a grant of $100 million and the European Investment Bank (EIB) along with the State Bank of India (SBI) has disbursed ₹200 million towards the development of various projects.

The investment by the private sector has been on the rise in recent years; however, there is a lot of potential that remains untapped to reach the targeted investment requirement. It is paramount that the private sector becomes an active player in the green finance market and brings its agility and commercial soundness on the table. Given that our country is a growing low-carbon environment market, the rise of private investment is inevitable. The need of the hour is to make the investment climate more conducive and financially viable in order to offer the private sector enough reasons to invest in this area. There is also a need to develop a formal definition of 'green' which will bring further clarity in terms of the projects which can be classified as green and ensure better understanding across sectors, and hence better targeting of finances.

Regulatory bodies in India like the RBI and Securities and Exchange Board of India (SEBI) have done an excellent job to reinforce investor confidence in the nascent green finance market, in an effort to broaden it. The RBI has augmented the extent of partial credit enhancement provided by banks to 50 per cent from 20 per cent of the bond size issue. The FICCI-UNEP Inquiry Report on Delivering a Sustainable Financial System in India[1], released in 2016, highlighted how India is introducing innovative approaches to attract private capital for green assets—and outlined a number of key steps to deepen this process. One of the recommendations included developing a sustainable capital market strategy building on SEBI's guidelines to further scale up the green bonds market. SEBI has published its official green bond guidelines and requirement for Indian issuers in 2017[2], placing India among an exclusive class of countries which have developed national-level guidelines.

---

[1] http://unepinquiry.org/wp-content/uploads/2016/04/Delivering_a_Sustainable_Financial_System_in_India.pdf
[2] https://www.sebi.gov.in/legal/circulars/may-2017/disclosure-requirements-for-issuance-and-listing-of-green-debt-securities_34988.html

INSTRUMENTS FOR GREEN FINANCE

FICCI has also launched the India Green Bonds Council in collaboration with Climate Bonds Initiative to facilitate the development of a green bonds market in India. By carrying out policy advocacy, market education and investors–issuer interface, the aim of the Council is to bring together investors, (public/private sector) banks, insurance industry, companies and other local market actors in a neutral forum to provide inputs for drafting the 'National Blueprint' that will be used to facilitate and promote the issuance of green bonds.

What we now certainly require is a mitigation of financing problems known to us. Green bonds have come up as a key financing mechanism to finance environment-friendly businesses and assets. The global green bond market has grown rapidly and was at $235 billion in cumulative issuance till the first half of 2017. India's green bond market, measured in cumulative issuance as of 2017, pegged at $3.2 billion, is the eighth largest green bond market in the world. Following the success of green bonds, there is now a need to innovate and create alternative instruments for green finance.

In that respect, The Climate Finance Lab, which works globally for mobilizing finance for climate change, has been effective in streamlining such innovative alternate instruments for green finance. Other than Sustainable Energy Bonds, the lab provides a host of other alternate services. It provides matchmaking services between investors and municipalities to help clear the project pipeline. It also undertakes an underwriting exercise, insurance activities and foreign exchange hedging for green financing activities. The Indian arm of the Lab was launched in 2015 as a public–private initiative to provide solutions to the financing challenges to investment in green infrastructure in India.

The idea of developing a 'Green Bank' is also an attractive proposition. The Indian Renewable Energy Development Agency (IREDA) has announced plans for itself to become India's first green bank. Establishing a dedicated green bank entails several benefits. Firstly, it can attract a pool of resources specifically directed towards environmentally sustainable projects at a low cost. It can help private banks execute initial transactions for clean energy projects via their risk mitigating products. A credible green bank may also help attract much required foreign private capital. Its specialization and core competence will help finance smaller projects more smoothly. Eventually, a green bank should help develop a deep market for green finance via demand aggregation.

GREEN FINANCE EXPERIENCES

Multilateral organizations have also taken a lead role towards green financing and are expected to continue to remain a prominent investor group in the coming years. The UNFCCC has established a Green Climate Fund (GCF), set up by all 194 countries that are a member to the UN. The ADB too has set up the Climate Change Fund (CCF). Together, both funds have received pledges to the tune of $15 billion in a span of two years. Such funds pay attention to the needs of society today, especially in the developing and the least developed countries (LDCs). Such funds will help open new markets for green investments.

The NDB established by the five BRICS nations in 2015 for promoting infrastructure and sustainable development projects in emerging economies has also set a strong example. It was expected to extend its cumulative loan portfolio of about $4.5 billion by the end of 2017 towards green projects, out of which $1.5 billion was so extended in 2016. Given the important role played by these organizations, it is imperative to provide further encouragement to them by bringing in suitable changes in the methodologies used for rating their green finance instruments, with such instruments attracting better ranking as compared to other options. The investors should also be acknowledged and given higher ratings for their green financing efforts. India should also closely follow the green finance experiences of Western counterparts and learn from them.

The BCS which was hosted by FICCI in September 2017 in New Delhi brought together various stakeholders from different countries on a common platform to share their ideas on climate change. Representatives from both business and government, from various advanced economies like the UK, France and Wales advocated measures taken in their respective home countries for lowering their carbon footprints. These success stories provide lessons for developing industrial countries like India where significant proportion of energy requirements is still met through coal. India meets around 60 per cent of its energy requirements through coal, while 16 per cent is met through renewables.

The future path for green finance needs to be more pronounced and well defined. All stakeholders have well realized that climate change is upon us and we need to act and act quickly at that. Global developments, as well as developments in India in this direction clearly show that we are taking the right steps to mitigate it, but a lot more needs to be done. If we continue on this path, I am confident India will be able to achieve the goals she has set for herself, well in time.

Financing Green Energy in a Power-starved Country

India, with economic growth above 7 per cent, is probably the only major economy, apart from China, experiencing consumption-driven growth today. And rising consumption means rising demand for services, products, infrastructure and various commodities, including energy-producing items such as coal and oil.

India is an energy-starved nation where huge power plants are lying idle because of unavailability of fuel, especially coal. Just imagine the big boost power production would receive once we resolve the fuel supply conundrum. On the one hand, we want to fill the power deficit by mining coal to fire the Ultra Mega Power Projects (UMPPs), on the other hand, we want a cleaner environment for our future generations. The answer lies in better energy efficiency and renewables forming a higher percentage of our energy mix.

There are multiple reasons for the restricted or muted growth of renewable or green energy in India and those have been widely debated across platforms—from the government to the private sector and from energy groups to voluntary organizations. While everyone agrees, at least in principle, to the need to promote green energy, the scarcity of funds for building huge green energy capacities has been ignored.

When the world is moving towards implementing the Paris Agreement to build a global architecture post 2020 on climate change, India is setting its own vision and domestic ambitions by scaling up its clean energy targets, planning 100 smart cities and setting strong energy efficiency measures.

With the Narendra Modi government taking concrete steps to build a 'green energy' culture in the country, with targets of 100 GW of solar energy and 60 GW for wind by 2022, the time has come for us to build a financial support system for this sector to proliferate. This is critical in the Indian scenario because our banks are already stretched financing traditional power projects, leaving hardly any room for green energy projects.

While we need massive funding for the green agenda, the paradigm shift would come from creating a sustainable framework for the financial sector that would change the rules of the game for financial institutions and create the appeal for financing the 'green'.

I sincerely believe that the banking sector should provide only the early-stage finance with the takeout being through the issue of bonds in the capital markets. There also has to be a lot of emphasis on debt and equity products. Efforts must be taken to establish a market-based mechanism triggering private capital investments into protection of the environment.

This is not easy. While it needs a stronger will from the financiers, it would also require suitable regulations from the government, RBI, SEBI and the insurance regulator, Insurance Regulatory and Development Authority (IRDA).

Let us look at some numbers to get an idea of the green energy sector. The global 'green market' is estimated to reach $203 billion by 2021 and stakeholders across the world are working towards meeting its financial needs. The international green bonds market saw $34 billion of issuance last year, growing rapidly from only $10 billion a year before that. Green bonds and yield companies are providing innovative routes for financing 'green' projects.

The market for such innovative products must be developed with urgency. The green asset class is emerging as the raison d'être of future corporations. Therefore, it is time the financial sector realigns itself towards the green asset class and the green economic agenda of the future.

TOWARDS A GREEN AGENDA

Coming back to India, we have already set some ambitious targets for green energy development. This includes doubling of the existing renewable energy capacity to 175 GW by 2022, supported by raising funds through tax-free bonds. Wind energy leads both installed capacity and capacity addition among all renewable technologies in India, contributing more than 65 per cent of the total renewable installed capacity in the country.

Interestingly, solar energy, in a country like India with ample sunlight, has not picked up as it should have. And the challenges are clear—lack of capital as well as institutional support and poor management.

Apart from involving foreign investors and development banks, the Indian government should look at tapping the domestic market to raise funds. The Export-Import (EXIM) Bank of India recently raised $500 million through a green bonds issue, while Yes Bank has raised $150 million through a similar tool. The government had requested eight entities to raise funds through green bonds. India's largest power producer, NTPC, has announced plans to raise at least $500 million through green bonds to fund its own solar power capacity addition.

On the other hand, KPMG has agreed to provide assurance services on an annual basis to ensure that the funds being raised are used as per green bond principles.

Let us look at the example of Tamil Nadu, a state which has emerged as a major hub for renewable energy over the past decade. More than one-third of the state's energy comes from renewable sources, such as wind and solar energy. Tamil Nadu achieved its current level of renewable energy development largely through government subsidies, such as depreciation benefits and generation-based incentive mechanisms. The state also has many large electricity consumers, especially heavy industries and the manufacturing sector, which are using cost-effective and reliable renewable energy.

So if this kind of expansion of green energy is to be done across the country, we have to have the financial systems in place. Restricting excessive investments in polluting sectors and incentivizing private investments in green industries could be an effective idea. We should also expedite the development of a green finance system for directing private investments to green industries and projects. Can we get our pension funds and insurance companies to hold some green investments as is the case with global funds in these sectors? Green ratings, green stock indices and mandatory disclosures can help steer funds into green industries. We also need to set norms for our banks on the style of the Equator Principles which ensure that minimum standards for environmentally sound projects are set and that companies which do not meet this standard cannot access finance.

The business case for financing of sustainability has to be created. The UNEP Inquiry on Designing a Sustainable Financial System along with FICCI has set up an India Advisory Council to propose practical solutions for creating a framework for sustainable financing. In its interim report, the Council creates the argument for developing a sustainability-oriented market framework that would eventually catalyze capital flow towards clean energy and other sustainable development priorities.

However, there are some bottlenecks that impede the flow of finance into the sustainability sector which require attention. We need bankable projects, and credit enhancement products will help make such projects more readily financeable. Building stronger green development financing institutions such as the Indian Renewable Energy Development Agency by increasing its bank book size, garnering additional lines of credit and long tenor financing and making it well positioned to deploy global green funding, would enhance financial flows to clean energy.

India has an extensive regulatory framework for the financial sector. Indian banking regulations and RBI directives hold the power to direct credit to specific sectors and further influence interest rates, exposure limits, incentives, security and other terms and conditions of lending to various sectors. We should direct priority sector lending policy towards funding of 'sustainable' businesses by allowing them to qualify for priority sector lending.

Today, it is mandatory for Indian banks to direct 40 per cent of their lending towards the priority sectors and funding renewables has been identified as priority lending. Bank loans up to ₹150 million meant for solar- and biomass-based power generators, windmills, micro-hydel plants and non-conventional energy based public utilities, namely street lighting systems and remote village electrification, qualify as priority sector lending. For household use, the loan limit has been set at ₹1 million per borrower. These limits would be reviewed in due course. One idea worth exploring here is clubbing loans for renewable energy with mortgage loans. One, it will incentivize home owners to go for renewable energy options for the home, and two, it will ensure that the loan amount for individual borrowers is collateralized through the mortgage.

The next step is to build a sustainable financial ecosystem, which believes in the green agenda. Can we have certain tax incentives for companies investing in the green market? This will ensure financing is effectively channelled to those that are ahead of the greening threshold. We could channel global climate finance into existing public expenditures on climate mitigation and adaptation to augment the ongoing effort under central and state government budgets.

Some other measures could be to allow external commercial borrowing for green projects by exempting withholding tax, replacement of construction finance and refinancing, and lowering net worth criterion for hedging of green projects. This will help smaller projects to avail international finance.

Renewable Purchase Obligation (RPO) mandated by Electricity Act, 2003, coupled with the more recent market-based Renewable Energy Certificate (REC) mechanism should push up the demand for renewable power. This should also help lower the cost of financing renewable energy projects while making way for shorter payback gestation.

Sustained growth in demand for renewable power is a prerequisite for commercial banks to make greater strides from running pilots to financing large-scale projects—which is what we all should strive for in the best interest of our future generations.

Financing India's Urban Development Pathway

There is already a major transformation taking place around us. Although we remain one of the last major economies in the world yet to urbanize, that is changing fast. Some estimates suggest that thirty Indians move from a rural to an urban area every minute. By 2050, India's urban population will nearly double to 814 million from around 410 million in 2015.

Our cities hold the key to our future prosperity. Our 100 largest cities already account for 43 per cent of India's GDP whilst accommodating just 16 per cent of the population. Over the next fifteen years, our cities will also account for 75 per cent of India's national income. They will also be home to the majority of new jobs created. In fact, research from the NCE showed that better, smarter urban growth could be an economic opportunity for India worth up to an additional 6 per cent of GDP by mid-century, with significant savings at the household level.

Ahead of us is the opportunity to ensure a better urban development pathway: cities where people can thrive, breathe, move safely and easily and be productive. This pathway requires investing substantially in urban infrastructure at a scale far greater than we have managed thus far, and doing so smartly. It is estimated, for instance, that as much as 80 per cent of the urban infrastructure that will be needed in India in 2050 has yet to be built.

TOWARDS COMPACT, COORDINATED AND CONNECTED URBAN DEVELOPMENT

Getting the right kind of infrastructure in place—that which ensures more compact, coordinated and connected urban development—will be key to ensuring that we avoid locking in unsustainable urbanization and instead help lift millions out of poverty. Recent research has shown that our current poorly planned, sprawling, unconnected pattern of urbanization could impose an estimated cost of ₹2–12 lakh crore ($330 billion to $1.8 trillion) by 2050. This includes increased costs of providing public infrastructure and services, transportation costs, traffic casualties, traffic congestion, air pollution and health risks. The costs could be much higher when factoring in increased road and parking capital requirements which could be upwards of ₹4 lakh crore ($600 billion) per annum by 2050, plus other costs such as the value of displaced agricultural farmland. The costs of providing public infrastructure and services are also likely to be 10–30 per cent higher in more sprawled, automobile-dependent neighbourhoods compared with more compact, connected locations.

The urban infrastructure investment gap in India is estimated to be around ₹6.4 lakh crore ($1 trillion) in 2011 prices, from 2012–31, although this number may be even higher. Meeting this investment need will require India to raise, carefully steer and blend new finance, from both the public and the private sectors, towards compact, connected and coordinated urban development. The Government of India should take this opportunity to look at the various tools available to enable investments in sustainable infrastructure and urban development and drive reforms that can provide an enabling environment for private capital to flow in at the scale required. Limited public finance needs to be carefully used in ways that will leverage private investment.

OUR CITIES TODAY
Building the cities of the future necessitates a hard look at the cities of today. We need to do better on many fronts—such as air pollution, congestion and sanitation—if we are to reap the benefits of urban development better. Our cities pose unique challenges that require tailored responses. While we cannot simply lift solutions that may have worked in other countries and apply them to India, we can certainly draw relevant insights, particularly in how they may have unlocked financing for sustainable urban infrastructure.

Prime among the challenges we face is alarming levels of air pollution—fourteen of the world's thirty most polluted cities are in India. Outdoor air pollution in Indian cities is estimated to cause around 1.1 million premature deaths per year. In Delhi, for example, the severe local air pollution in November 2017 forced the government to order all schools to be closed in the city.

High levels of air pollution are often linked to urban transport with up to 75 per cent of urban air pollutants coming from fuel combustion in motorized vehicles. Our roads are quite literally deadly with the highest number of fatalities in the world—more than 1.46 lakh deaths annually. In addition to the human cost, congestion also places an economic burden on commuters. In Delhi, traffic congestion costs averaged ₹4.91 per kilometre for cars and ₹9.83 per kilometre for buses during peak periods.

Quick fixes often inspired by efforts in other cities, such as odd–even traffic restrictions that worked for Beijing ahead of the 2008 Olympics, rarely offer a lasting solution. Rather, a comprehensive package of urban development should include policies on congestion charging and regulated parking fees, while enhancing public transit options in order to make them overall a more attractive alternative for city dwellers.

An especially unique challenge for our cities is that they are extremely dense. Homes in Mumbai have only about 30 square feet per person, less than a quarter of what is available in urban China. Sprawl in India means something quite different than in regions such as the United States. It is better understood as a low density of built-up floor space per unit of land area, combined with severe overcrowding per unit of built-up area. Addressing sprawl in India therefore will require a greater emphasis on 'appropriate' or 'good' density combined with adequate provision of accessible and well-connected infrastructure and services.

URBAN FINANCE REFORM: A BOOST TO EXISTING URBAN POLICIES
A number of government initiatives are already focusing on improving the quality of urban development. Smart Cities is an urban renewal and retrofitting programme to develop 100 cities with a focus on core infrastructure services such as water, sanitation and solid waste management, efficient urban mobility and public transportation, affordable housing for the poor, power supply, robust IT connectivity, governance, safety and security, health and education and sustainable urban environment. AMRUT ensures that every household has access to water and sewage connections, and encourages increase in green and open spaces and public or non-motorized transport options, particularly with support from the private sector. It

largely replaces the earlier Jawaharlal Nehru National Urban Renewal Mission (JNNURM) programme aimed to make urban infrastructure and service delivery mechanisms more efficient, increase community participation and improve accountability to citizens. SBM's main objective is to reduce or eliminate open defecation through the construction of individual, cluster and community toilets. Finally, Pradhan Mantri Awas Yojana-Urban (PMAY-U) seeks to provide affordable housing to the urban poor.

Of particular note in the current challenge is the role played by ULBs, the third institutional tier of urban government, after the central and state authorities. ULBs often lack the necessary combination of technical capacity and fiscal autonomy required for urban finance reform, but could be key to helping deliver infrastructure services, planning, regulation, welfare, health and safety. Improving their governance, capacity and expanding their remit to manage urban finance would provide a strong complement to these flagship urban initiatives and help to deliver sustainable urban development.

A holistic urban finance policy package that raises, steers and blends finance from public and private sources, supported by strong institutional leadership and a productive enabling environment that enhances local capacity and improves governance can support the goals of these initiatives and help unleash our full urban potential. Some steps to achieving this are as follows:

Raising Finance

There are broadly three sources of potential finance for urban development. First, own-source revenue, such as tax revenues (property tax yields the majority of ULB income, but there are others, such as for lighting, vehicles, advertisement, etc. which vary greatly across the country); and non-tax revenues (for instance, user charges for services such as water). Research has shown that reformed property tax mechanisms could increase revenues by as much as 71 per cent, but many ULBs lack the capacity to administer and collect both taxes and other revenues effectively and efficiently.

Second, grants from state governments to ULBs (including, for instance, funding available through AMRUT or SBM to improve basic infrastructure or solid waste management, as well as fixed and performance-based grant allocations, and potential GST-derived revenue) are substantial sources of revenue. However, only nine states received 100 per cent of the basic grant recommended by the Fourteenth Finance Commission, whilst nine states received 50 per cent or less in 2015–16.

And third, debt financing, where wide variations across the country exist because ULBs lack the sound financial footing to borrow, and even where borrowing is possible, the restrictions or approvals required by states often prevent it. Notable exceptions are large, financially powerful cities such as Ahmedabad and Bengaluru, and the pooled financing mechanisms being deployed by smaller ULBs in Tamil Nadu and Karnataka.

To increase the ability of ULBs to raise finance, we might consider ways to build their capacity to collect own-source revenues and demonstrate municipal creditworthiness in order to attract private investment.

- Improve the efficiency of property tax collection, such as by establishing Property Tax Boards that would set up statewide methods for property valuation and assessment (as has been proposed by the Thirteenth Finance Commission) and keep the collection of tax as well as decisions about spending at the ULB level. We should also explore innovative solutions for registering properties, including through Geographic Information Systems (GIS), perhaps using the Smart Cities style competition approach so cities that demonstrate improvements in certain areas may receive financial support for more expensive tools.

- Make service delivery financially sustainable, for instance, by unbundling user charges and subsidies. This approach would enable ULBs to set charges for users at a level that would sustainably finance utility investments over the long-term, whilst separately supporting the poor urban groups that otherwise cannot afford to access such services.
- Make fiscal decentralization more efficient and empower ULBs to develop capacities. The GST could provide a unique opportunity to offer some of the financing needed and support ULBs—the 25–30 per cent share recommended by the MoUD could be used to fund their infrastructure requirements as well as develop the capacity of ULBs, and would be a valuable compensation for the revenues they give up as part of the GST's creation.
- Improving ULBs' creditworthiness to enable access to debt financing, by creating a good enabling environment. This could mean, for instance, expanding the credit ratings developed by the MoUD under Smart Cities and AMRUT, currently covering ninety-four out of 500 cities, and developing Ministry of Finance (MoF) guidelines for ULBs on the issuance of tax-free bonds. States can undertake similar supporting efforts as a complement to these central efforts.

Steering Finance

Meeting our urban infrastructure investment needs will require us to steer large-scale finance effectively into compact, coordinated and connected urban development. Some steering measures already exist, such as the Energy Conservation Building Code, which encourages investment in energy-efficient building approaches and is already being implemented in twenty-two states; and the Indian Corporate Average Fuel Consumption Standard which seeks to encourage investment in more efficient vehicles and the tax on coal, which steers finance towards sustainable investment via the Clean Energy Fund.

To ensure finance is appropriately steered into sustainable urban development, we should explore the potential of mechanisms that capture increased urban land values to finance sustainable infrastructure. These include:

- Refreshing regulations and incentives, whereby those that encourage 'sprawl' are removed or amended (such as minimum parking requirements, setback requirements and maximum building heights) and replaced with policies that encourage dense, sustainable developments (such as access to public transport, minimum sanitation requirements to reduce pollution and encouraging density by establishing maximum floor space). Other policy levers, such as parking fees and congestion charges can also encourage the use of public transport and reduce the congestion that, as above, hampers our overall development.
- Expanding the Land Value Capture Financing policy (implemented so far only at the central level from April 2017) to the state and ULB level to capture the increases in land value created by their investments

in transport and infrastructure. Businesses already recognize the value of developed urban land—that explains its high cost—and governments can capitalize on this. But to really be successful, such measures should be complemented with clear master plans that steer development towards more compact urban form, along with social safeguards that demonstrate structured, long-term urban growth.

- Using competition to incentivize sustainable urban infrastructure investment. Smart Cities demonstrated that competition was a powerful motivator for ULBs to act. An annual ranking of cities based on key sustainability measures (similar to the Swachh Survekshan survey that ranks over seventy cities based upon their cleanliness, expanded to cover air pollution, energy efficiency, congestion and public transport provision) could act as an incentive for cities to invest their own resources in sustainable urban infrastructure.
- Engaging with other national or international funds. One of the key recommendations of the recent UNEP-FICCI report on sustainable finance in India which I chaired, was to better leverage international funds, like the GCF. The GCF or the $500 million UK-India Infrastructure Fund can effectively provide cheaper loans for sustainable urban infrastructure with the central government playing a key role by helping support sustainable projects.

Blending Finance

During my time at HSBC, we started to engage fully with issues of sustainability. Acting on multiple fronts—green investments, reducing our own carbon footprint, seeking out sustainable business opportunities and investing in communities—helped make our business more sustainable while also making a perceptible difference to communities in which we operate. Banks can ensure we do not fund companies that do not comply with norms set by Indian regulatory authorities and can therefore push companies to comply. The Indian Banks' Association (IBA) issued guidelines on responsible banking in 2015 which is a good start for ensuring that all banks include sustainability in their lending decisions.

The private sector has a major role to play in meeting our urban development needs. The 12th Five Year Plan envisaged almost 50 per cent of the total investment coming from the private sector. PPPs were also outlined as key objectives in JNNURM. In 2011, over half of all investment in PPP projects in developing countries occurred in India. But despite the increase in private investment in infrastructure, PPPs in India have a number of drawbacks. Contract enforcement and lengthy litigation processes—as much as 30 per cent more time than the average in South Asia—pose major challenges that deter private sector investment.

As a result, private sector actors who do invest can often renegotiate terms midway and gain additional advantages at the expense of the public good. Issues around land acquisition have also stalled PPPs and any measures to address this must take into account the needs of vulnerable urban communities. Investors are also wary of some urban infrastructure projects where the returns are harder to demonstrate because of public good attributes (water, as noted above, a clear example of this), and where the regulatory landscape is crowded with multiple government agencies overlapping.

In order to blend finance at the scale required, we may wish to build investor confidence in PPPs by addressing the effectiveness of contract enforcement. We must, therefore,

- Address the challenges around contract enforcement and dispute resolution, including through effective, quick, and low-cost alternative resolution mechanisms such as forming an independent body or mechanism that can facilitate mediation between disputing parties and without further burdening the court system. Special courts specifically set up to deal with contract enforcement can also speed up resolutions with PPPs and make them more attractive to investors.
- Other measures, including pooled financing mechanisms, may also be explored to meet the needs of smaller projects with state governments or donor institutions playing a role in guaranteeing risk or enhancing creditworthiness. Replicating, for instance, the Tamil Nadu Water and Sanitation Pooled Fund—a successful pooled finance mechanism that enabled ULBs to access private finance that would otherwise have been inaccessible—could be a good complement to the bigger blended model offered by PPPs.

TURNING AMBITION INTO ACTION

These recommendations are well-suited to attract the financing we need for India's urban transition and transform the urban financing landscape. As a country with an established financial system and governance framework as well as globally competitive urban centres, we are well-positioned to both leverage advanced financial instruments that draw upon the resources of a strong private sector and better utilize the capabilities of strong and mature public institutions.

At the same time, we must stay aware of how financing

urban infrastructure will impact the poorest and most vulnerable—to make sure that investment is well-governed and equitable. For instance, if user charges for key services go up (which they should to ensure long-term investments in maintaining the infrastructure), they should be accompanied by strong and inclusive social protection measures. In terms of land acquisition, governments should work with current owners and occupants to find mutually satisfactory arrangements. Low-income and other marginalized urban residents should not be displaced unnecessarily or without adequate compensation to create formal commercial space or high-income residences.

How India finances its urban expansion today could be of great interest to other countries seeking to do the same in the future. Some of the measures to evaluate success relate specifically to ULBs—for instance, the sum and proportion of own-source revenue they generate, their creditworthiness and total lending to them and how those loans perform. We should also assess capital flows into sustainable urban infrastructure as well as how these are funded. These measures can complement assessments of more traditional urban and environmental indicators, such as access rates to housing, infrastructure and services or air pollution and emissions, etc.

A more holistic approach to financing our urban future is available to us, and essential if we are to realize our full economic potential.

Global Finance and Global Warming

Since 2008, when the global financial crisis nearly brought down the world economy, financial reform has been among the top items on policymakers' agendas. But, as leaders move from fixing the problems of the past to positioning the financial system for the future, they must also grapple with new threats to its stability, particularly those stemming from climate change.

That is why a growing number of governments, regulators, standard-setters and market actors are starting to incorporate rules concerning sustainability into the financial system. In Brazil, the central bank views the integration of environmental and social factors into risk management as a way to strengthen resilience. In countries like Singapore and South Africa, companies listed on the stock market are obligated to disclose their environmental and social performance, a requirement that investors and regulators increasingly view as essential to the efficient functioning of financial markets.

Initiatives like these might once have been regarded as part of a peripheral 'green' niche. Today, they are considered central to the operation of the financial system. In Bangladesh, the central bank's efforts to support economic development include low-cost refinancing for banks lending to projects that meet goals for renewables, energy efficiency or waste management. In the United Kingdom, the Bank of England is currently evaluating the implications of climate change for the insurance sector as part of its core mandate to oversee the safety and soundness of financial institutions.

In China, annual investment in green industry could reach $320 billion in the next five years, with the government able to provide only 10–15 per cent of the total. In order to prevent a funding shortfall, the People's Bank of China has recently produced a report with the UNEP setting out a comprehensive set of recommendations for establishing China's 'green financial system'.

In India, FICCI has established a new 'green bond' working group to explore how the country's debt markets can respond to the challenge of financing smart infrastructure. And recent regulatory changes hold out considerable potential for listed investment trusts to deploy capital for clean energy.

So far, such measures affect only a small fraction of the $305 trillion in assets held by banks, investors, financial institutions and individuals in the global financial system. But they are set to be applied more broadly as financiers and regulators alike recognize the full consequences of environmental dislocation.

Those consequences already are severe. In 116 of 140 countries assessed by UNEP, the stock of natural capital that underpins value creation is in decline. The human and economic costs of continued high-carbon growth include severe health impacts, growing disruption to infrastructure and water and food security, as well as increasing market volatility, most notably in developing countries. This damage will become worse, with risks becoming unmanageable if emissions of GHGs are not reduced to net zero levels between 2055 and 2070.

As the threat from climate change becomes more evident, financing the response to its impact will become increasingly important. Developed countries have committed to mobilize $100 billion in annual financial flows to developing countries by 2020, but much more is needed.

Above all, it is essential to place the financing challenge posed by climate change within the broader context of the green economy and sustainable development. The task for those charged with governing the financial system is to enable the orderly transition from high- to low-carbon investments and from vulnerable to resilient assets. According to the NCE initiative, $89 trillion will be spent on global infrastructure investment by 2030—with an additional $4.1 trillion needed to make it low-carbon and resilient.

To mobilize the required capital, policymakers will need to harness the power of the financial system. The scope of risk management will need to be expanded, so that long-term sustainability and risks from climate change are included in prudential rules for banking, insurance and investment. New 'green banks' can help to bring in funding from debt and equity markets. Transparency will have to be improved through better corporate reporting and enhanced disclosure from financial institutions. Financial professionals' skills and incentives will have to be retooled and revised to reflect these new priorities.

Promising avenues for international cooperation are now opening up. For example, the G20 finance ministers and central bank governors have asked the Financial Stability Board to explore how the financial sector could address climate issues. Actions such as these will not only strengthen climate security, they will also contribute to a more efficient, effective and resilient financial system.

Building the BRICS of the Paris Agreement

The year 2015 was a testament to the worldwide, broad-reaching support for action to tackle climate change. There are many important steps being taken, particularly by some of the biggest developing economies.

The Paris Agreement entered into force on 4 November 2016, with all countries finding common ground in a vision of a zero-carbon future. Over 100 countries have ratified it, including China, India, Brazil and South Africa. This was incredibly fast, setting records for speed for international agreements. And more countries continue to join.

The first meeting of the parties to the agreement took place in Morocco in November 2016. The COP 22 conference in Marrakech was an important agenda setter for the next few years, providing the framework for countries to ramp up their ambition. The BRICS countries have already stepped up their leadership post-Paris to reduce emissions and promote sustainable development. Supported by UNEP, there have already been efforts to incorporate sustainability factors into the rules that govern the financial system. Now however, we need to accelerate what so far has been a 'quiet revolution'. The need of the hour, now more than ever, is for these countries to really show leadership in acting on climate.

The BRICS' NDB has also been an important mover, making clean energy finance a vital part of its mandate. It launched its first four investments in April 2016, worth $811 million, all for renewable energy projects. Then, in July, it issued 3 billion renminbi ($450 billion) of green bonds in China's interbank bond market. The fact that the NDB's first debt issue was designed to finance clean energy sends an important signal.

On a country-by-country level, the BRICS are making great strides to integrate the meaning of the Paris Agreement into their economic strategies. China unveiled its newest Five Year Plan, mapping out the country's strategy for economic and social development. The 2016–20 plan places great emphasis on the greening of the economy. In fact, the math suggests that China will likely over-deliver on its climate commitments for 2020.

The Chinese government acted quickly, and in the first eight months of 2016, China shut down 150 million tonnes of coal mining capacity, with another 100 million tonnes of capacity scheduled to close in another year. It is no wonder China does not need coal anymore—it has become a giant in the sustainable energy economy. In 2016, China invested around $100 billion in renewable energy, which is 36 per cent of the global total. In 2017, it planned to implement the world's largest emissions trading scheme, expanding its seven pilot trading systems to the national level.

Next door, we in India have exciting clean energy dreams of our own. India has committed to expand its renewable capacity to 175 GW by 2022, from 34 GW in 2015. In another encouraging move, India has been phasing out subsidies to diesel and petrol. In March 2016, the government doubled the tax on coal, lignite and peat. It uses the proceeds to fund clean energy. Programmes towards enhancing energy efficiency such as use of LED bulbs are well under way.

2016 was an important year for Brazil too, not least because the Olympics were hosted in Rio de Janeiro. Rio didn't squander the opportunity to highlight climate change—in the opening ceremony, it broadcast a video about global warming to a global audience of billions, boosting the profile of the issue to an extent it rarely receives. For its part, Brazil aims to increase its proportion of non-hydro renewables from less than 10 per cent today to 23 per cent by 2030.

The other BRICS countries are taking positive steps as well. South Africa plans to become the first African country to put in place a carbon tax. It has built Africa's first solar-powered airport. Russia, meanwhile, has begun to reform some fossil fuel subsidies and tax breaks. Its commitments to the Paris Agreement may not have been particularly ambitious, but its energy strategy aims to generate 4.5 per cent of electricity from renewables by 2020, from less than 1 per cent today.

The meetings in Morocco have laid the groundwork for the next few years of climate action, and can help raise ambition to make sure we all do our part in limiting global warming to safe levels. That is a good goal for us to ensure the well-being of citizens from all countries.

Developing countries played a massive role in getting the Agreement to the finish line. Now they should take the lead in making the low-carbon economy a reality.

The Case for a BRICS Rating Agency

The BRICS Business Council has over the years deliberated on several key economic issues critical for the development of the grouping. One of the areas that has attracted much attention, particularly amongst members of the Financial Services Working Group, is that of a 'BRICS Rating Agency'. Most of the BRICS countries require huge infrastructure and ensuing finance. We need to aid the process of funding coming into our countries and use this finance effectively to build green infrastructure. A ratings process that helps in these directions will benefit not only BRICS countries but also the world at large.

The concept of establishing a new rating agency for the emerging markets was first proposed during the Russian presidency in 2015 and has been discussed extensively amongst group members since then. After considerable deliberations on the subject, the BRICS Business Council included the idea among its key proposals at the BRICS Summit held in Goa, India, in 2016. We also received approval of the BRICS leaders' to work more on this concept.

While sceptics may scoff at this idea, just as they did when the NDB was proposed, a deeper study of the issue at hand shows that there are solid reasons for us to move ahead with this proposal. And for this reason, following the Goa Declaration, the BRICS Business Council formed an expert group to deeply evaluate the need and feasibility of setting up the BRICS Rating Agency based on market principles. Members with extensive experience in the financial sector and capital markets from all five BRICS countries were invited to join this group, which is presently working with a set terms of reference to examine the various aspects related to the rating agency such as viability, regulatory framework, ownership, governance and management structure of the entity.

The group shared its interim report with the BRICS Business Council ahead of the 2017 Summit meeting in Xiamen, China, and its findings and suggestions indicate that there is a clear need for a rating agency for BRICS and other emerging countries. In fact, a survey done amongst potential users of credit ratings across the five BRICS countries also showed that there is strong demand for a new rating agency that would offer a much better understanding of the emerging markets. Let me explain this in some detail.

Currently, ratings are assigned based on either national or global rating scales. These have limitations. National scale ratings of a domestic rating agency in one country are not comparable with those from a different country; ratings on the global scale usually get restricted by the country's sovereign rating, leading to bunching of the ratings at the lower end of the scale. Hence, an alternate rating agency is required to provide ratings on a broader emerging market scale that is an intermediate between the national and global rating scales, one which will offer a sharper differentiation in credit ratings for the benefit of investors focused on emerging markets. This will help investors in getting a much better idea about the credit quality of the issuers and thus help them take a more informed decision on allocation of funds within these geographies.

The investors and financial market players surveyed also shared some of the aspects which would play a critical role towards the success of the BRICS Rating Agency. These include experienced management and strong corporate governance; extent of coverage provided by the rating agency both in terms of the number of borrowers covered and types of financial instruments rated; ability to provide differentiation of risk across borrowers in emerging markets; and accreditation by regulators and independence from political influence.

Members of the expert group have taken note of these expectations and based on the detailed discussions held made a set of recommendations. Let me take you through some of these and the rationale behind the same.

First, the ownership and management structure of the BRICS Rating Agency. The expert group is of the opinion that the rating agency to be free from any political bias (and from a single country) should have a diversified and independent shareholding structure. For the BRICS Rating Agency, we may therefore consider dividing shareholding amongst national development banks of BRICS nations, multilateral bodies and other public and private financial institutions. The NDB could also be a stakeholder to ensure long-term stability and independence from a single sovereign. The rating agency could also consider receiving support from established credit rating agencies present in the BRICS nations as this could enable the entity to gain faster acceptance among the issuers.

Second is the regulatory framework for the agency. Governance norms and regulatory framework lend credibility to and enhance acceptability of a rating agency among investors. Hence, members of the expert group feel that it is important that the location where the BRICS Rating Agency is based offers a robust regulatory framework in line with global standards. Moreover, factors such as the size and reach of capital markets and the presence of a well-developed domestic credit rating industry will improve visibility of the BRICS Rating Agency. All these factors should be considered while evaluating the country of establishment. Preferably, the rating agency should be established in one of the BRICS countries and a 'Governing Charter' can be formed and signed by all the founding member countries to govern the agency.

The third aspect which has been studied by the group is the scope of services to be offered by this new rating agency. The discussion here has been closely linked to establishing the reputation of the BRICS Rating Agency. And for this, it has been suggested that the proposed rating agency must adopt a methodology that is based on fundamentals and predicated upon a rigorous rating model which should be fair and transparent as well. The ratings should clearly delineate the differences in the credit risk profile of issuers relative to the risks in their specific markets. Further, the rating agency should be empaneled with the central banks of the BRICS nations and relevant credit rating regulators of the five BRICS countries as this would improve its acceptability amongst financial institutions lending in foreign currency. With this kind of an approach and set-up, the agency could initially focus its ratings on (a) corporate sector debt, (b) public sector debt and (c) financial institutions.

For the business model, the rating agency should adopt an 'Issuer Pays' model which is widely accepted and commercially successful worldwide. Such a model offers a number of advantages. All ratings under this model are available publicly free of cost and secondly, it allows the rating agency to access relevant information from the issuer on a regular basis, which is critical for proper assessment of creditworthiness of issuers and thus assigning correct ratings to financial instruments issued by the issuers.

The proposed rating agency can initially serve the BRICS nations but its services could gradually be extended to other emerging markets as well. However, before extending its scope of activities to other countries, the rating agency should identify the final users of the ratings in each of the countries and understand their respective needs. Amongst all sectors, it is the infrastructure sector that requires maximum attention and funding in emerging markets. And as we see a gradual increase in the participation of emerging market funds in infrastructure build-outs in emerging economies, a credit rating agency whose evaluation is reliable, transparent and independent will proactively carry forward the development agenda of the emerging economies. Thus, forward movement on the proposal for a BRICS Rating Agency along with simultaneous discussions with institutions like NDB, ADB, Asian Infrastructure Investment Bank and the World Bank will be a very encouraging step. We need investors across the board, including capital market players, to endorse the move and commit to using its services.

Chasing Green Jobs

India is keen to attempt to work towards a low-carbon emission pathway while simultaneously endeavouring to meet all the developmental challenges. The Intended Nationally Determined Contribution (INDC) is taking forward the Prime Minister's vision of a sustainable lifestyle and climate justice to protect the poor and vulnerable from adverse impacts of climate change. India's INDC centres around its policies and programmes on promotion of clean energy, especially renewable energy, enhancement of energy efficiency, development of less carbon intensive and resilient urban centres, promotion of waste-to-wealth, safe, smart and sustainable green transportation network, abatement of pollution and India's efforts to enhance carbon sink through creation of forest and tree cover. India, at COP 21 in Paris, declared a voluntary goal of reducing the emission intensity of its GDP by 33–35 per cent, over 2005 levels, by 2030. India has adopted several ambitious measures for clean and renewable energy, energy efficiency in various sectors of industries, achieving lower emission intensity in the automobile and transport sector, non-fossil based electricity generation and building sector based on energy conservation. Thrust on renewable energy, promotion of clean energy, enhancing energy efficiency, developing climate-resilient urban centres and sustainable green transport network are some of the measures for achieving this goal.

The Indian INDC brings a huge responsibility on the country and also opportunities for green business and skilled manpower requirement towards creation of a green economy. The green economy is no longer an aspirational phrase, but a compelling way of sustainable living. It is driven by widely accepted citizens' concerns over environment, climate change, water and waste and has an articulated roadmap in the INDC which forms part of the Paris Agreement.

Green jobs are jobs that contribute to preserve or restore the environment, whether in traditional sectors such as manufacturing and construction, or in new, emerging green sectors such as renewable energy and energy efficiency or services such as audit and rating of green activities.

At the enterprise level, green jobs can produce goods or provide services that benefit the environment, for example green buildings or clean transportation. However, these green outputs (products and services) are not always based on green production processes and technologies. Therefore, green jobs can also be distinguished by their contribution to more environmentally friendly processes. For example, green jobs can reduce water consumption or improve recycling systems. Yet, these jobs defined through production processes do not necessarily produce environmental goods or services. They are central to sustainable development and respond to the global challenges of environmental protection, economic development and social inclusion. Greening of enterprises, workplace practices and the labour market as a whole can be achieved by engaging governments, workers and employers as active agents of change. These efforts create employment opportunities, enhance resource efficiency and build low-carbon sustainable societies.

GROWTH POTENTIAL OF GREEN BUSINESSES

The scope of green jobs covers the entire gamut of 'green businesses'—renewable energy, energy storage, green construction, green transportation, carbon sinks, solid waste management, water management and e-waste management. Hence, they would have a pan-India impact.

Highlighting the job creation opportunities that a scaled-up clean energy market offers in India, analysis by the Natural Resources Defence Council (NRDC) and the Council on Energy, Environment and Water (CEEW) estimates that solar photovoltaic (PV) projects built in India between 2011 and 2014 created approximately 24,000 full-time equivalent (FTE) jobs—solely from commissioned projects currently producing electricity. The wind sector has created about 45,000 FTE jobs so far, according to government estimates. Despite limited data, solar and wind renewable energy is estimated to have created nearly 70,000 FTE jobs in India so far. If India achieves its target of 100 GW of installed solar energy by 2022, as many as one million FTE jobs could be created. Approximately 1,83,500 FTE jobs would be generated if India were to reach its target of installing 60 GW of wind energy capacity by 2022. Looking ahead, solar and wind companies in India can support the clean energy market by reporting the jobs created by their projects.

These days there is also a renewed buzz and energy around the sanitation sector in India. Undoubtedly, the SBM has catalyzed the conversation around the issue, right from the streets to the boardrooms of corporate India. Through platforms like the ISC, discussions on the questions of financing, corporate engagement, technology solutions and the need to address sanitation across the value chain of BUMT have been brought to the forefront. However, what is missing is the conversation around the economy of jobs, as it relates to the sanitation sector. That is, when talking about the supply side, there is a need to deliberate on not just infrastructure creation, but also job creation. In order to achieve the end goal of sustainable sanitation that includes ODF++ beyond 2019, the creation of a workforce to support this agenda is critical. This renewed buzz in the sector should therefore be used as an opportunity to shine light on the need for a workforce to support the sustainable sanitation agenda as well as an opportunity for employment generation in the country.

Skill development and job creation in the sanitation sector must go hand in hand. With the massive scale of toilet construction under way across the country, the training and certification of existing masons and plumbers is critical to ensure that these toilets that are being built are of the highest standard, technologically appropriate and sustainable over a long period of time. In light of this, there are already certain corporates that have identified the need to create a skilled manpower for sustainable sanitation. One such initiative is the Kohler Plumbing Academy (KPA). It offers an excellent example of corporates working towards creating a shared value wherein social and business concerns are addressed. The Academy was created to address the challenges of meeting the demand–supply gap. It strives to train more plumbers and meet that demand through a structured professional education with the vision that it will create social entrepreneurs who will further facilitate employment through training across India. It is interesting to note that the programme also focused on promoting plumbing as a glorified job, to instil a sense of pride in the target group. Simultaneously, they also created opportunities for the trained workforce to be absorbed by the industry.

The NSDC is a PPP in India, set up in 2009. Its primary objective is enhancing, supporting and coordinating private sector initiatives for skill development. NSDC aims to skill/ upskill 150 million people in India by 2022, mainly through private sector initiatives and providing viability gap funding. It works through the formation of Sector Skill Councils (SSCs) which are national partnership organizations bringing together all the stakeholders—industry, labour and the academia for the purpose of workforce development for particular industry sectors. Studies conducted by the Skill Council for Green Jobs indicate that sixty-five million jobs will need to be created by 2030. Interestingly, the largest potential being in sectors of waste management and water management, each of which constitute 30 per cent of the forecast for total jobs. Green construction would constitute 17 per cent and green transport about 12 per cent of the forecast. Renewable energy and carbon sinks would constitute the balance 11 per cent of the forecast for total jobs. It is, hence, necessary to adopt a holistic perspective towards the green economy, with the circular economy being an embedded principle.

Skilling for green jobs has the advantage that there are few legacy issues to address. Hence, the focus can be on forecasting evolution in green businesses, which have exponential growth potential. This growth is driven by rapid advancements in solar/storage and biochemical/thermochemical technologies, as well as digital technology. These trends facilitate adoption of a distributed architecture for production and consumption of products and services that are constituents of a green economy.

MAXIMIZING SKILL DEVELOPMENT

There are several sectors where skill development would be required. In the green buildings sector, it is important that future green building programmes and projects are established with strategies to address skills issues, including appropriate training components. The importance of labour is particularly apparent in retrofitting of existing buildings, where labour costs make up a very large proportion of the total costs (often well over 50 per cent, in the case of wall insulation). In this situation, there are clear economic benefits in maximizing labour productivity and eliminating the need to replace substandard work. Labour productivity and quality of work are both related closely to skills quality.

In the green transport sector, the shift from a diesel-based mass transport system to CNG involves engine modifications, and requires an increase in two types of employment—filling station attendants and mechanics. In India, currently the skills gap is more for mechanics, with a shortage of authorized service centres. Building technical capacity will be essential. The demand for biofuel and bio CNG trained mechanics will rise dramatically, as will the need for quality training institutions. In the absence of formal training, non-formal training arrangements are evident, where mechanics trained in diesel engines pick up the knowledge relating to CNG engines through on-the-job training as they work alongside formally trained mechanics.

ECONOMY OF JOBS IN SANITATION
This is a huge opportunity waiting to be tapped in the sanitation sector. At present, there is no separate SSC looking at sanitation. The two SSCs which are working on sanitation related job roles are the Construction SSC and Plumbing SSC. What we need going forward is to leverage the various platforms built around skill development to bring the conversation around the economy of jobs in the sanitation sector to the forefront. This includes platforms like the Pradhan Mantri Kaushal Vikas Yojana (PMKVY) that was launched more than two years ago with the

aim of giving as many as 2.4 million young Indians industry-relevant training with an elaborate certification scheme. The target for skilling is also aligned to demand from other flagship programmes launched such as the SBM.

The ISC's core philosophy is around BUMT. These are the four steps towards sustainable sanitation, representative of the entire value chain. The potential of creating livelihood avenues and entry points at each step are tremendous. For example, the waste management stream of the sustainable sanitation value chain has a great potential of generating highly specific skilled jobs. The process of scientific waste management requires skilled and trained manpower to handle various sub-activities such as waste collection and transportation, recycling, treatment and disposal. Such opportunities need to be brought to the forefront, formalized and leveraged. Furthermore, according to corporates like Kohler, the sanitation industry in India is sized at about ₹5,000 crore and needs approximately 3,00,000 trained and qualified plumbers. The need is growing at a rate of 12 per cent every year. But the current plumber strength is less than half the need from the industry. Such insights and practical knowledge can only be brought to the surface by bringing together multiple stakeholders like corporates, implementation partners and multiple ministries within the government to discuss and deliberate on a way forward.

ENTREPRENEURSHIP SKILLS AND MODELS

Skilling entrepreneurs who can set up businesses to provide much-needed public services is critical. Often many of these services can have a pay-for-use model so that they are profitable ventures. In Kampala, Uganda, 10 per cent of faecal sludge is collected by small entrepreneurs who have become millionaires. However, this required the local municipality to work with an

NGO to ensure the collection of sludge is done in a clean, humane way in sealed containers and wheelbarrows, taken to vans which deliver to a plant on the outskirts of the city. The sludge is treated there and then used as fertilizer in neighbouring cotton fields. In Dharavi, entrepreneurs maintain community toilets and have made reasonable profits in a model developed by SPARC, an NGO which works with the local community and municipality. We will need businesses that set up small FSTPs, run them and maintain them to a high standard. We need to help establish these models and then have entrepreneurs run these as their livelihood.

At the ISC, we believe that the opportunity around the economy of jobs in the sanitation sector is immense and yet to be tapped. India has about 30,000,000 million unemployed youth, educated yet unemployed due to the absence of any professional qualification or vocational training. To move forward, the ISC is working to spur the discussion around job creation and economic models in the sector, leveraging skill development platforms as well as the start-up wave engulfing India. In order to ensure sustainability, there is a dire need to address the issue of the economy of jobs in the sanitation sector to support the overall agenda to make India ODF by 2019 and beyond.

Therefore, creation of a skilled workforce is absolutely essential for the green economy. Educational institutions have to collaborate with the government as well as industry to develop course curricula with specific needs to address such issues. Vocational courses in such aspects may also help a great deal. Numerous universities in the country now offer courses on environmental studies, environmental management, conservation biology, natural resources management and power management, but specific skill sets need to be developed to address technical fields like transportation, construction, waste and water facilities designing. A sound academic base of environmental professionals in the country along with right opportunities in the green sector will lead to our country's sustainable growth.

Women

Women's Empowerment: Nurturing the Ecosystem

Gender inequality has remained a pressing social and economic issue globally and more so in emerging economies like India. A recent study conducted by McKinsey has shown that if the gender gap in occupation can be reduced or if women participation in workforce is made equal to that of men, it could add as high as $28 trillion or 26 per cent more to the global GDP by 2025. The report further suggests that India will be one of the highest beneficiaries with an additional boost of 60 per cent to its economy by 2025. In such a scenario, it is imperative that the issue of gender equality, particularly at the workplace, is treated with utmost importance.

India's status in terms of gender equality as compared to other global economies, particularly the Nordic and western European countries, can be considered as dismal. This is evident from the Global Gender Gap Report 2015 published by the World Economic Forum, which shows that Iceland, Norway, Finland, Sweden and Ireland are the top five economies in the world with least gender gap. These countries have achieved 100 per cent education for women at all levels, attained healthy life expectancy at birth irrespective of sex and have been able to provide quality job opportunities for educated women workforce with proper work–life balance.

India's performance has not been significant as compared to Latin American and Caribbean countries, which have witnessed the largest increase in women participation in workforce between 2014 and 2015. In sub-Saharan Africa, women's participation in workforce has increased by 3.2 per cent which has helped in narrowing the gender gap further. Contrary to these regions, India has widened its gender gap in workforce participation by 11 per cent in 2015. This is despite the fact that India has witnessed rapid economic growth in the last decade.

There is also wide disparity in wages paid to men vis-à-vis women in India, both in private as well as in public sector for formal or informal activities or work. The National Service Scheme (NSS) data of 2011 show that men on an average earn almost four times more salary than women in both formal and informal sectors in urban and rural India. Financial inclusion of women in India is also quite dismal with only 26 per cent of women having formal accounts with banks or other financial institutions compared to 46 per cent of men. Also, only 15 per cent women entrepreneurs in India have access to formal financial credit. A Goldman Sachs report in 2014 pointed out that the rate of rejection of SME loans for women in India is 2.5 times more than that of their male counterparts.

While getting a job or starting a business, women have severe issues in India, getting a job however does not guarantee a problem-free environment for them. Studies show that most workplaces in India, both public or private, do not take enough care of women aspirations in terms of providing them growth opportunities, promotions and offering them executive positions. A recent report highlighted that woman representation into entry-level management position is only 24 per cent, and it is even lower at 14 per cent for the top positions—women hold only 7.7 per cent of board seats and just 2.7 per cent of the board chairs. Out of the 323 executive director positions on the Bombay Stock Exchange (BSE), just eight are held by women. The industries with highest percentage of women on board include media, technology and telecommunications.

Also, many studies point towards the fact that the workplace environment and infrastructure in India are not women-friendly. Facilities such as company-sponsored parental leaves, access to childcare and healthcare facilities within office, flexibility in shift patterns, etc. are not present in most cases. Globally, it is found that such facilities have actually encouraged more women to participate in the workforce and have increased their productivity. These facilities help women take care of several of their other responsibilities, including their role as the primary caregiver in the household. Absence of such facilities leads to excessive mental stress among female employees and many are forced to leave jobs with increasing family size.

COUNTERING THESE PROBLEMS

Education

Education is one of the most important determinants of boosting women participation in economy, having a U-shaped relationship; as education level among women rises, their workforce participation declines at first and then increases, especially among highly educated women. In India, between 2001 and 2014, the Gross Enrolment Ratio (GER; ratio of number of individuals who are actually enrolled in schools and the number of children who are of the corresponding school enrolment age) for girls in higher education improved from

6.7 to 19.8. The GER for girls in primary classes was very high at 100.6 in 2014. The government's 'Beti Bachao Beti Padhao' initiative has been instrumental in achieving these high levels in primary education and also in keeping India on track of its MDG of attaining universal education. However, the GER for girls in senior secondary and higher education is relatively much lower at 49.1 and 19.8, respectively, showing the impact of increasing dropouts which can be attributed to factors like societal pressure, lack of funding and resources, increasing responsibilities like taking care of younger siblings or early marriage. Safety of schoolgoing adolescent girls is also a concern in India. In fact, India's 2014 average GER in higher education for both male and female students at 20 was much below the global average of 30. Therefore, India must strive to achieve the global average level in the next five to ten years, by taking measures to encourage more girls to complete their higher education. Developed countries like the US and in Europe have been able to attract more females in higher education by providing greater responsibilities to them in education management system. These women have acted as role models for girls, thereby helping in increasing enrolment of women in higher education. Bangladesh presents a good learning in this regard. The country has partnered with International Labour Organization's (ILO's) Technical and Vocational Education and Training (TVET) Reform Project to boost female participation in the TVET projects by providing female-friendly environment in both training centres and workplaces. Initiatives taken include provision of skills training for workers in the informal economy and establishing an adequate data management system to capture sex-disaggregated data on TVET.

Financial Inclusion

With several programmes such as the Pradhan Mantri Jan-Dhan Yojana and Pradhan Mantri Bima Suraksha Yojana, India's financial inclusion has progressed considerably. However, gender inclusion still remains a concern. While the account density of females has more than tripled between 2006 and 2015, it is still far lower than males. Given this scenario, banks have to make special efforts to step up account opening for females. Financial service providers should realize the importance of women in their customer base and design policies in a way so as to provide maximum benefit to them. There should be women-specific financial services and products to target maximum customer base. Providing loans through microfinance and self-help groups is not sufficient to encourage women entrepreneurship in the country. There is a need to train these women in business and finance management as well. Financial education programmes increase self-confidence of women and positively impact their income generating ability.

Developed countries have dedicated programmes for women with the objective of making them financially educated. Austria's Department of Social Service has a financial and consumer education training programme targeted at women returning to work after a gap. The Swedish Enforcement Authority found in 2010 that women's proportion of debt in respect to short-term credit has increased. To resolve this, it engaged in a preventive communication project called 'Women and the Economy' targeted at women aged above 26 years.

Given the fact that loan applications of women entrepreneurs often get rejected in India due to absence of tangible assets which can be mortgaged, appropriate policies should be framed to give asset/property rights to women, including land rights.

Similar provision has been introduced for rural women in Brazil under its National Documentation Programme, which can be an important learning for India.

Skill-based Training

Along with financial education, it is important that women receive specific apprenticeship training and vocational courses from a young age. Lao People's Democratic Republic is gearing up to provide stipends to girls for training in traditionally male trades—automotive and mechanical repair, carpentry, furniture making, electronics, plumbing and metalwork. Incentives will be given to training providers to enrol girls in trade courses; and wage subsidies will be provided to enterprises for employing girls. The project further sets female quotas (20 per cent) for training in three priority skill areas—construction, furniture making and automotive and mechanical repair.

India should use such global experiences and plan to devise similar programmes here as well. FICCI Ladies Organisation (FLO), established with the prime objective of women empowerment in India, through its business consultancy cell in Delhi has been providing consultancy services to potential female entrepreneurs. The organization not only helps women set up units, but also provides appropriate guidance during the running of these units.

It is a well-known fact that the provision of skills, some financial training and microfinance enable rural women to earn through livelihoods, substantially changing their position in the communities in which they live and the way they spend on their families' welfare. We need more organizations like Self Employed Women's Association (SEWA)—nurturing self-employment of our women.

Workplace Infrastructure

Provision of important facilities such as flexible parental leaves, childcare facilities such as day care and health programmes are very important for encouraging women and increasing their productivity. Several developed and developing countries have enacted policies aiming towards making office infrastructure and services more women-friendly. India has also passed the Maternity Benefit (Amendment) Bill, 2016, which has introduced a number of women-friendly provisions, including raising the maternity leave from twelve weeks to twenty-six weeks for working women, which is a welcome move. The Bill has also made it mandatory for firms with fifty employees to have crèches individually or in arrangement with a few firms within a prescribed distance. These initiatives would be helpful in attracting and retaining more women in the workforce.

Sweden's high female workforce participation has been attributed to generous and flexible parental leave policy, high coverage of childcare, job guarantees and eligibility for availing flexible working hours. More than 90 per cent of companies in Germany and Sweden allow flexible working. A growing number of firms are learning to divide the working week in new ways, judging staff on annual rather than weekly hours, allowing them to work nine days a fortnight, letting them come in early or late and allowing husbands and wives to share jobs. In Netherlands, with the introduction of breakdown in barriers between full-time and part-time work contracts with well-compensated parental leaves, part-time workers benefitted from the same hourly wages, social security coverage, employment protection and rules as full-time workers. This increased female participation from 35 per cent in 1980 to 80 per cent in 2008. In 2004, Brazil adopted a National Plan for Women policies to address specific needs of mothers, including healthcare during pregnancy, as well as childcare and education.

India's own programme, Mahatma Gandhi National Rural Employment Guarantee Act (MGNREGA) is a very important example too. The Act provides for facilities such as childcare at worksites to reduce barriers to women's participation. Other aspects such as the fact that work has to be provided within five kilometres of the applicant's residence, works very well for women employed under the scheme.

Women in Corporate Governance

Studies show that countries that have greater representation of women in public life are known to have reduced levels of socio-economic inequality. Women tend to advocate and prioritize investments which have societal implications relating to family life, health and education. With growing confidence, they vote sensibly which strengthens democratic processes. India has initiated some steps in this direction—there is already one-third reservation for women at the Panchayat level. Studies show that while initially their husbands use them as a front, after ten to fifteen years, they begin to assume the responsibilities of the role and fulfil these effectively. The Women Reservation Bill remains a disappointment. For encouraging more women in politics, the following steps can be considered:

- Taking up the Women's Reservation Bill can be an effective step towards increasing women participation in parliament;
- Sensitizing political parties to include more women legislators and adding more gender-related issues in their agendas; and

- Running of the National Women's Machinery should be free of any political influence and should be given a clear mandate.

Studies also indicate that closing of gender gaps has a positive impact on the results of companies as well. A linkage has been shown to exist between having more women directors and effective corporate sustainability programmes. Having women at the leadership level has been seen to ensure a better perspective with regard to consumers and caters to a broader array of stakeholder needs. Policy measures encouraging more women participants in the boards of publicly traded companies have been found to be effective in many countries. For instance, in Norway, prescribing a quota of 40 per cent for women representation in boards of publicly traded companies resulted in an increase in women representation on board from 9 per cent in 2003 to 40 per cent in 2008. India has also taken initiatives in this direction. According to the Companies Act, 2013, every listed company and every other public company in India having paid-up share capital of ₹100 crore or more or turnover of ₹300 crore or more, are required to appoint at least one woman director on the board. Similarly, in 2014, SEBI mandated to include at least one woman director on the boards of the listed companies starting from April 2015. This is a good beginning but we need to move beyond the one in due course and better governed companies are beginning to show the way.

Sanitation and Other Basic Facilities

Lack of proper sanitation facilities is another area of concern for women in India. It is estimated that nearly 300 million women and girls in India defecate in the open. This makes women vulnerable to different types of crimes including rape and murder, which take place when they step out early in the morning or late in the evening to defecate. Lack of sanitation facilities also leads to severe health problems for women, including infections of the urinary tract, kidneys and sexual organs. This issue, therefore, needs urgent attention. The Swachh Bharat Abhiyan launched by Prime Minister Narendra Modi is a step in the right direction. However, what is critical is to ensure that the various goals set under this programme are achieved in a timely fashion, which is essential to resolve this issue.

Water scarcity is also a major problem in semi-urban and rural areas of India. Since women are primary caregivers responsible for making food and drinking water to the family members, they are the most affected by this problem. Reports suggest that many girls in the country had to drop out of school as they are required to travel to faraway places to bring water. This problem, therefore, needs proper attention and measures be taken to ensure that adequate supply of water be provided to the areas facing water scarcity.

Safety for Women

India is considered to be an unsafe place for women. Strict measures are required to make the country a safer place. Sexual harassment policies for offices as prescribed under the Vishakha Guidelines can help in preventing such incidents from taking place at the workplaces. However, it is important to ensure that all companies adopt these guidelines for their respective offices as well as victims getting justice in a speedy manner. It is also important to encourage women to come forward and register complaints against perpetrators.

Road and communication infrastructures should be made secure for women. India can learn from Vietnam's Ho Chi

Minh City Mass Rapid Transit Program which has been designed to promote safe and secure physical mobility for women. Promoting gender equality by introducing programmes in schools and parenting programmes to sensitize families on the cases of women and child sexual and mental abuse and maltreatment should be implemented.

There is a need to sensitize the police as victims of crime often approach them first. In India, police apathy towards handling crimes against women is a well-known fact which needs to change. All offences against women should be given due importance including cases of molestation and eve-teasing, and perpetrators should be handled with strictness irrespective of their social and political background. For victims of crime, strong institutions for counselling should be available. A mechanism of acknowledging the special vulnerability of rape victims should start from the first act of registering the complaint, and the criminal judicial system should provide adequate support to ensure that the victim has equal access to justice. There is also a need to sensitize the justice system in India. Court proceedings for such women can be painful and lengthy. Also, the long time taken to give justice is seen as a problem for the victims.

Landmark Initiatives

FICCI has undertaken various initiatives to nurture the ecosystem for promotion of women's empowerment and demonstrates the role associations and industry bodies can play. It had set up FLO, a national body, with the prime objective of women empowerment through the promotion of entrepreneurship and managerial excellence. FLO members have also taken a pledge to help create a climate of respect towards women and ensure safety for women.

FICCI is working towards a target that women and young leaders should constitute at least 30 per cent of its Sectoral Committees' composition. It has also encouraged its members through its guidelines encompassing the Vishakha Guidelines regarding sexual harassment at the workplace. FICCI's member companies are also advised to have a detailed diversity and inclusiveness policy and determine how they intend to keep decisions free of bias and based only on meritocracy. To ensure gender diversity on corporate boards, the FICCI Centre for Corporate Governance has undertaken the 'Women on Corporate Boards' programme. FLO has published a ready reckoner on 'Laws affecting women in India' to fill the knowledge gap about women's rights in the country. With a view to increase the employment of women as support staff in schools in India, FLO launched the FLO Women Empowerment (WE) in Education initiative in 2015. To address the issue of lack of sanitation facilities in girls' schools, FLO has tied up with Sulabh International to provide toilets in schools. As part of the project, schools are provided with toilets and sanitation facilities. FICCI Socio Economic Development Foundation (SEDF) is also working actively towards this objective. FICCI is also the secretariat to the ISC, which has been set up to bring all organizations working in sanitation onto a common platform with particular focus on women's needs.

India has come a long way in women's participation in the economy. However, there are still some gaps to bridge. Taking care of some of the above issues will go a long way towards encouraging more women in the workforce.

Rural Indian Women: Sowing the Seeds of Social Transformation

When you come across terms like 'women's liberation', 'gender equality' or glass ceiling', in all probability you will visualize a young, modern woman holding her own in a competitive corporate environment, maintaining a balance between her work and life, and being a woman of substance. While this image may not be entirely misplaced, there is a large section of women in India who are gradually breaking the mould and transforming society in a quiet way.

Despite a massive migration towards urban areas, 68 per cent of the Indian population still resides in rural areas. This means close to seven out of ten women live in rural areas. It signifies the importance of empowering these women to bring about a positive social change.

A woman's role in the predominantly rural sectors like agriculture, forestry, animal husbandry and education cannot be overlooked. Today, as women begin to augment household incomes, the attitude towards working women is changing. Slowly, but surely, women are finding their place in the workforce.

Women working in the unorganized sector, who are self-employed or homemakers, may not contribute to the GDP numbers, but their contributions have been considerable. Most of their jobs are manual, unskilled and low-paying, but they provide a huge boost to a typical rural household income. Moreover, these women play a major role in shaping and uplifting their family, community, as well as society as a whole.

Whether it is about tilling the fields, running a cottage industry or teaching in an Anganwadi, rural women have the power to transform society in their own unique way. I believe that rural women today have a stronger voice in the village owing to a number of changes taking place in the past few years. Some of these changes are visible through instances like those mentioned below.

GREEN CONSCIENCE KEEPERS
Women have played a pivotal role in creating a green and clean environment. The HSBC partnered with a local NGO in Gujarat to work in five villages of Dahod district for the enhancement of rural livelihood of 400 families through environment-friendly development interventions aiming to mitigate climate change at a micro level. The project entailed innovative work in the area of climate change, land regeneration, natural resource-based livelihood and income generation, carbon sequestration and overall improvement in the local economy.

The activities were a combination of land development by increasing green cover in wasteland and farm bunds; income for rural communities through agroforestry (completely managed by tribal women from the area); vermicomposting (development of organic manure using earthworms); fruit orchards, horticulture, floriculture and vegetable cultivation.

These initiatives resulted in the planting of 1,60,000 trees on wastelands, 400 trees by each of the 400 farmers, and the tree cover restoring and improving the ecosystem. A hundred vermicompost units produced substantial organic manure to enrich land quality and ensure much higher agriculture yield (one vermiculture pit produces 6 quintals compost which could fetch about ₹300 per quintal). Also, 100 plots were raised by 100 farmers, which after a gestation period helped each farmer to earn around ₹40,000 per year from one orchard of half an acre.

Owing to these initiatives taken by women from the five villages, there was measurable increase in income levels of farmer families, 106.7 hectares of wastelands covered by a total of 1.6 lakh trees matured into carbon sink; 195.73 tonne carbon dioxide per annum was sequestered. Overall, 400 rural livelihoods benefitted from a project leading to positive climatic change.

In fact, women's contribution has been significant in agricultural and forestry sectors. A sizeable number of rural women are employed in the forests for conservation of green cover. There is also a vital link between effective water management and rural women's role in water conservation. The role of women comes into the foreground as in villages only women fetch water for everyday use. To increase awareness about water-related issues, HSBC joined hands with SHAR in Maharashtra. With HSBC's support, women in different areas of the state created self-help groups (SHGs) to work on innovative structures, such as springs cordoning, bund, underground bund, recharge pond trenches and rooftop rainwater harvesting which helped them gain better access to water resources.

MANAGING THE HOUSEHOLD BALANCE SHEET

According the Ministry of Labour and Employment, the participation of working women in India has risen steadily since 1981. Contrary to popular belief, there has been a considerable rise in the number of working women in rural areas, especially when compared with urban cities. The data from the census of 2001 show an increase of a mere 1.81 per cent between 1991 and 2001 in the cities. The rise was higher in rural areas in the same period; it rose from 27.20 per cent to 30.98 per cent, an increase of approximately 4 per cent. These figures suggest that women are increasingly joining the workforce, thus increasing the dependence of rural families on a woman's income.

According to a study, the income generation under Development of Women and Children in Rural Areas (DWCRA) has seen an overall increase in the average income of a rural household. The sum total of a rural family's income from all trades adds up to ₹2,989 per annum. So if a rural family is engaged in dhoti weaving, it earns ₹7,200 and ₹600 more for undertaking activities like beekeeping, juice- or fibre-making and preparing shampoo or vermillion. With such increase in earnings, 146 out of 200 village respondents could easily manage to rise above the poverty line. A major portion of this income was contributed by rural women. Similarly, their urban counterparts have witnessed substantial increase in their household income. In 2013, it was $31.1 trillion.

With more women now earning, it is interesting to analyse their share in household spending. In India, a working woman's household spending is around 44 per cent, whereas the highest number is that of the US, with 73 per cent of the household spending being controlled by women. As a whole, both their earning and spending capacity has brought about a notable change in the society.

INCREASING INDEPENDENCE

I believe that once a woman steps out to earn her livelihood, she becomes independent, not just economically but psychologically too. She gains better control over the family's finances and the overall household and thus acquires stronger decision-making powers. Furthermore, research suggests that women contribute a larger share of their earnings for the welfare of their family. A woman usually spends her resources for the betterment of her children, their education, healthcare, recreation and other basic needs. This kind of spending pattern creates a healthy impact on society as a whole. This tendency emerges even stronger when compared with an average rural man who may have different priorities in life, which may not always be beneficial to his family. Due to this factor, the independence and earning capacity of rural women is directly proportional to the transformation of society.

MICROENTERPRISE AND MICROFINANCE

Credit provided by various banks and MFIs has changed the lives of several women all over the world. In fact, many countries in Latin America and Asia have numerous MFIs catering to the needs of millions of entrepreneurial women, helping them become self-employed. I am in favour of creation of microfinance opportunities in rural areas in our country as it provides women with the requisite financial resources to start their own venture. This, in turn, helps them gain confidence to step out into the world and create their own identity.

I can't help but mention a feisty lady, who managed to hold her own despite the odds stacked against her. Gauriben, a barely-literate woman from a small village in Gujarat, a young mother of two, lost her husband. In a flash, the responsibility of her entire family landed on her shoulders. In such circumstances, making ends meet was a challenge. To add to her woes, her community had a parochial and conservative approach towards widows, one of which was preventing them from gaining employment. Also, if she could muster the courage to venture out to work, there was no support system to take care of her children.

Despite these obstacles, Gauriben married her daughter into a good family and learnt to read and write. Her hard work paid off—she is now the proud owner of a small house and a patch of fertile land, which she tills. Today Gauriben is a village leader in SEWA and a role model for many women. If microfinance opportunities succeed in creating more women like Gauriben every day, it will go a long way in placing rural areas on the fast track of economic progress.

BUSINESS OPPORTUNITIES

The introduction and development of local industries has enabled rural women to participate in agricultural and local production activities like handicrafts. Their contribution to the household income has had a positive impact on the GDP numbers too. To aid them further, the Government of India is promoting micro entrepreneurship in all states. This has led to the emergence of many women entrepreneurs in remote, smaller regions. Moreover, these entrepreneurial initiatives have successfully contributed to the GDP, which in turn, has supported the Indian economy to a large extent. This impact is driven by the need for rural women to employ their entrepreneurial skills in different work areas and gain employment. A need for this is strongly felt among women from the villages, as they are in search of an additional source of income. The economic impact of this new class of employed female population from rural areas is not only empowering women, but also guiding society in the right direction.

CORPORATE PARTNERSHIPS

The Fabindia model is a great example of how a corporate can encourage employment at a rural level. By facilitating the setting up of SHG-owned companies, from whom Fabindia buys and continuously provides training and design inputs, a win-win situation is developed where both the corporate and the women benefit.

For a number of years, my husband and I have been closely engaged with Grassroots Trading Network for Women, an NGO promoted by SEWA, which has helped SEWA members forge relationships with corporates. HUL trained these women to distribute and pack products. The rural distribution network of SEWA, RUDI, was born, and today members of SEWA earn money, distributing products across their villages, packing some of their own grown seeds and spices and earning money through their sales. ITC provided skills and inputs in order to enable these women to achieve norms, which allowed it to buy from them. This helped women to earn ₹2 per kg extra and introduced consolidation and other standards to a wide constituency. Other successful partnerships have included Usha Sewing Machines and Hero Bicycles.

HIGHER EDUCATION

With rising exposure and access to higher and better quality education opportunities, today's rural woman can make informed decisions. She can have a say in important financial matters and play an active role in the well-being of her family. The saying, 'Educate a woman, and you educate the whole family', is absolutely true in this context because a woman not only uses her knowledge for the betterment of her family, but also passes it on to her children and other members.

My personal experience with an organization called Mann Deshi Udyogini, a rural business school for women, and SEWA helped me understand the importance of education in the lives of these women. I feel that lack of education and skills was one of the foremost reasons keeping them from achieving financial independence. With a rise in the number of schools and vocational training centres, women in rural areas now have the opportunity to gain knowledge and hone their skills. As a result, we see women from smaller regions in India becoming engineers, doctors and even astronauts, which was unimaginable a few decades ago. India is hence fast catching up with developed nations in providing equal educational opportunities to girls and boys. Hopefully, we will soon be able to match the record of Southeast Asia, where an impressive 111 girls for every 100 boys are enrolled for higher education courses.

STEPS IN THE RIGHT DIRECTION

The Government of India has taken several measures to create employment avenues for rural women. Many government initiatives like the Right to Education (RTE) and MGNREGA are enabling the marginalized sections of society, including women in rural areas, to educate themselves and earn a living. An impressive 49.33 percentage of female population from across the country has benefitted by participating in the NREGA. By creating the right support systems for them, the government has played a key role in uplifting and empowering women.

In an effort to improve the employable skills of the rural female population, the Directorate General of Employment & Training (DGE&T) has set up one national and ten regional vocational training institutes. Through these centres, women from across India can attain skills as well as vocational training. Thousands of women have been trained in these institutes. Additionally, hundreds of women in different states are working in Industrial Training Institutes (ITI). Several women training wings have been constituted across the length and breadth of the country with systems for imparting basic training for skills development for women. The government's National Rural Health Mission (NRHM) and the regional women empowerment programmes have come as a boon.

TECHNOLOGY LIBERATES
History has shown that the biggest transformation in the lives of women happened with the advent of the washing machine, freeing women in the West from the drudgery of washing clothes and giving them time to pursue other interests and join the workforce. Technological progress has brought about a dramatic change in the way a job is executed. There has been a considerable increase in the female workforce with the necessary skill set to handle various types of mechanized tools and equipment. As a result, women are gradually foraying into unchartered territories. By gaining professional training on new technologies, women are continually improving their technological skills, due to which they are able to reduce manual working hours and strike a balance between work and life. Moreover, such training is pulling the rural population out of the unorganized sector into organized ones like IT, manufacturing, services and so on, which offer better chances of stable employment and professional growth.

An example of the need for automation is the case of the women in rural areas where the bucket irrigation system is prevalent which involves fetching water in pails manually for irrigation. Replacing the manual irrigation system with a tube well irrigation system will cut down a woman's physical effort to a great extent. She can then focus on other aspects of agricultural work and undertake multitasking. The concept of automation and technology is applicable in many other areas of agriculture like wheat grinding, weeding, tilling and so on. Modern techniques like wheat grinding machines, pesticides, herbicides and so on can help bring down a woman's effort considerably.

The 1960s saw a Green Revolution in the state of Punjab, which ushered immense prosperity throughout the region. Agricultural activities in the state are highly mechanized and the rural workforce, including women, hugely benefits from this. On the other hand, the state of Andhra Pradesh is extremely labour-intensive and is highly dependent on its female workforce for labour-intensive crops like cotton and groundnut. Automation and use of latest technology for irrigation in the state may help women increase productivity and achieve even better results than they are currently registering.

FOCUS ON WOMEN EMPOWERMENT

Local governing authorities like the panchayats have begun framing women-centric laws. As a result, more women like Gauriben face lesser resistance when they seek work opportunities in rural areas today. They are also able to achieve independence and generate income more easily. For example, there has been a consistent rise in the number of female representatives who participated in rural water projects. In fact, these numbers are higher than even their male counterparts. In addition, as the number of elected women representatives from different villages rises, their control over critical financial and social aspects also increases. They will have a stronger say in matters relating to family expenditures, which in turn will have a huge impact across various sectors like health, education and infrastructure. Rural women will also help in boosting the workforce in certain areas of the country where the working population is on the decline.

The role of rural women in India's growth story is significant. In addition to their stellar contribution to their family, community and village, these women have demonstrated their potential to change society and have paved the way for a modern India. As more Gauribens cross the threshold to discover their identity, they strengthen a major pillar of democracy—the rights of the nation's women and how they vote.

Reproduced with permission of Oxford University Press © Oxford University Press 2012

Women in the Sustainability Discourse: Time for Action

India's commitment to transition into a low-carbon economy provides the opportune moment to reflect on the successes and the lessons learnt from the MDGs era and the possible way forward for achieving the ambitious and inclusive agenda of SDGs. While the positive economic outcomes of this transition are well-articulated, this could also be transformative for the country in terms of addressing certain social impediments to its growth. Achieving substantial progress on this front could come by way of unlocking the potential of nearly half of our population—our women.

Research the world over suggests that when women contribute, economies grow. Yet the sociopolitical set-up in many developing countries does not provide a favourable environment for women to work to their full potential. On average, women in the labour market still earn 24 per cent less than men globally and a recent study shows that the gender pay gap in India is as high as 67 per cent, as a man on an average earns $167 compared with $100 by a woman.[1] India has amongst the lowest female labour force participation rates (LFPRs) in the world—well below what would be expected for its level of income and what is observed in neighbouring countries such as Bangladesh, Sri Lanka and Nepal.[2] A 2013 study conducted by the World Economic Forum highlighted a strong correlation between a country's gender gap and its national competitiveness, income and development.[3] A nation's competitiveness, the study assesses, in the long term, depends significantly on whether and how it educates and utilizes its women and men equally while enabling women to access the same rights, responsibilities and opportunities as men.

Empowering women and promoting gender equality is also crucial to accelerating sustainable development. This necessitates looking at them not just as a disadvantaged group (and hence the limited focus on 'mainstreaming' them), but also unlocking their expertise and experience for planning and decision-making. Much literature has focused on evidence-based studies on the relationship between women and their environment. The dialogue however often restricts itself to how women can be better impacted from development interventions as against how to utilize their expertise and potential by empowering them. As the UN Women position paper on the post-2015 development agenda[4] notes, women's empowerment and gender equality have a catalytic effect on the achievement of human development, good governance, sustained peace and

---

[1] https://newsroom.accenture.com/news/three-critical-accelerators-to-closing-the-gender-pay-gap-for-class-of-2020-accenture-research-finds.htm
[2] https://data.worldbank.org/indicator/SL.TLF.CACT.FE.ZS?
[3] http://reports.weforum.org/global-gender-gap-report-2015/the-case-for-gender-equality/
[4] http://www.unwomen.org/en/what-we-do/post-2015/un-women-position

harmonious dynamics between the environment and human populations.

SUSTAINABLE MANAGEMENT OF WATER AND SANITATION
As in much of the developing world, women and girls are the primary carriers of water in India while their participation in decision-making is limited. The water and sanitation sector over the years has seen an enhanced thrust on institutional reforms, including the increasing recognition of the bottom-up approach as against a top-down one. This stemmed from the need to enhance inclusiveness of all stakeholders, including women. Although a welcome departure from the earlier approaches, it did little to recognize the fact that women can contribute immensely to the decision-making process by their vast and unique experience, and therefore should not only be seen as potential beneficiaries. Studies have shown that involving women in design and planning stages can accrue multiple benefits to such projects. A World Bank evaluation of 122 water projects found that effectiveness of a project was six to seven times higher when women were involved.

An example of women leading the way can be best seen in the Indian context where a movement to make the country ODF by 2019 is currently under way. Women across the country have championed the cause of a Swachh Bharat at the grassroots and have played a leading role in the progress made so far. Their role has been particularly important in BCC and advocacy which lies at the centre of achieving any real success. They understand the health benefits of clean ways and sanitation once explained to them; they embrace the need for toilets for these reasons and the risks to their security as they otherwise go to fields under the cover of darkness. In India, women sarpanches have scripted success stories and made a difference in their villages, making them ODF by building toilets and creating awareness. There have been instances where women have forayed into the traditional masculine domains like masonry. The latest gender guidelines released by MDWS caution against stereotyping women by limiting their role as behaviour change agents. It also highlights the importance of making the IEC/BCC messaging gender sensitive as it was seen that some existing campaigns inadvertently propagated gender stereotypes.[5]

PROMOTING GREEN ECONOMY AND GREEN JOBS

Women are not only vulnerable to climate change but they are also effective actors or agents of change in relation to both mitigation and adaptation. In India, women are already engaging in green economic activities, working as forest stewards, farmers, natural resource managers and entrepreneurs (though this contribution is mostly unrecognized and undervalued). The need for integrating the gender dimension based on understanding the role of women vis-à-vis men while designing the intervention strategies has been emphasized in the past. Moving a step further, the present climate change dialogue is trying to integrate the development dimension by creating economic opportunities for sustainable growth. However, with respect to women, discussions have been mostly focused on how susceptible they are to climate change and how to use their knowledge and expertise for better programmatic outcomes. While this is critical, the discourse now needs to shift to ways of creating climate change mitigation and adaptation into avenues of economic empowerment of women.[6] One such

---

[5] http://www.mdws.gov.in/sites/default/files/Guidelines%20on%20Gender%20issues%20in%20Sanitation.pdf
[6] https://www.oecd.org/dac/gender-development/46975138.pdf

example was the task of making solar lamps that transformed tribal uneducated women into engineers in Udaipur, Rajasthan, turning them into green entrepreneurs. Similarly, in the Barefoot College in Tilonia, Rajasthan, women (often illiterate) are trained to manufacture and maintain solar panels in a six-month programme. Such efforts would need to be replicated and scaled up into viable business models to not only empower women as part of a green workforce, but also to turn them into green leaders.

Technology will play a key role in addressing the challenges of climate change. However, women's access to technology is a critical challenge due to social and cultural biases, inadequate technological infrastructure in rural areas, lower education levels, fear of or lack of interest in technology, along with a lack of disposable income to purchase technology services. In bridging these gaps great dividends can be earned.

GENDER IN POLICIES AND PROGRAMMES: EXAMPLES FROM WASH

The concept of 'gender' was conspicuously absent in the limited policy and plan documents on WASH until the beginning of the 1980s. The Water and Sanitation Decade (1980–90), which was the first instance when a comprehensive review and perspective was developed in the sector as a whole, brought in the concept of community participation and the need to look at the specific needs of women, though this was mostly restricted to rural areas and was limited to token statements like '…consulting and involving women at all stages of program and project planning…' Furthermore, the policies and plans were not backed by any specific guidelines and the concept of 'women's participation' remained a mere statement. For example, the Sanitation Rating Criteria, on the basis of which cities are to be declared totally sanitized or otherwise, does not have any specific indicator to assess inclusion and gender integration.

The process of ensuring participation and gender equity across sectors was further strengthened when a National Mission for Empowerment of Women was launched under the Ministry of Women and Child Development (MoWCD) in 2010. Access to health programmes, drinking water and sanitation and hygiene facilities for women is one of the six focus areas of the mission with convergence of schemes and strengthening the institutional framework for the same being one of the key strategies.

At the policy level, an absence of gender disaggregated data severely limits a sound understanding of women's roles and representation in decision-making. For instance, a study[7] on gender budgeting in water and sanitation with reference to the slums of Delhi concluded that although the union and most state governments have adopted the practice of gender budgeting and generate an annual Gender Budget Statement, water and sanitation is not specifically reflected in it. Further, the concerned water and sanitation departments and agencies at the union or state levels also do not provide gender disaggregated data, which made it difficult to assess the volume spent on women in the provision of water and sanitation.

Recently, the MDWS has released guidelines on gender issues in sanitation. The SBM (G) guidelines have a provision for 50 per cent representation by women in Village Water and Sanitation Committees (VWSC) and the state government and

---

[7]Kathryn Travers, Prabha Khosla and Suneeta Dhar (eds). 2011. 'Gender and Essential Services in Low-income Communities: Report on The Findings of the Action Research Project Women's Rights and Access to Water and Sanitation in Asian Cities', http://www.jagori.org/wp-content/uploads/2010/02/IDRC-Final-Report-on-the-Project-Findings-COMPRESSED2.pdf

local government bodies are encouraged to promote not only women's participation, but also their leadership in various decision-making institutions.

Sustainable development can only be achieved through long-term investments in economic, human and environmental capital. As India aims to transition on the path of sustainable development, the current policy focus on green economy provides a historic opportunity. Much would depend on how soon policymakers switch from regarding women as a disadvantaged group to powerful decision-makers with insights to drive strategies for a better future. The key priorities to make this a truly transformative agenda for women should inter alia include enhancing their role in private and public decision-making, enhancing access to finance, training and capacity building on technology and capturing gender disaggregated data for better policy interventions.

Learning from the challenges faced by women in the traditional non-green labour market would be critical to design better opportunities for women to access green and decent jobs in the green economy. Green economy by itself will not change the underlying anomalies like women's limited access to productive inputs like credit, land, technology, etc. After all, creating equal opportunities for women is not only a development imperative embedded in human rights, it is also critical to accelerating sustainable development.

Epilogue

Survive or Sink is a call to every one of us as Indians (whether citizen, industry, media or government) to play a role in our country's agenda for sanitation, pollution and the environment.

Today, there is an unprecedented buzz and energy around Swachh Bharat which needs to grow into a people's movement in order to achieve the ambitious target of India becoming ODF by 2019, and to sustain this thereafter. Such a people's movement would include those most directly benefitting from SBM as well as converting those who believe that they do not have a vested interest in sanitation. The truth is that it impacts us all—our health, the beauty around us and our pride. We need to ensure that the entire sanitation value chain is addressed, from building toilets to ensuring equitable access for every citizen, ensuring toilets are used and well maintained and that the human waste generated is safely treated and disposed of, ideally for productive use.

At the ISC, we work with multiple players like corporates on CSR funding and volunteering programmes, financial institutions driving sanitation financing and individuals with a passion to contribute to the mission. Ultimately, we need all hands on deck.

We need to encourage the role of corporates in water

efficiency and conservation through a water stewardship approach. In order to achieve this, there is a role for the government in promoting these policies; financial institutions in ensuring compliance and measuring water related risks; civil society in guaranteeing accountability of government and in educating consumers; and for consumers through their purchasing decisions and policy advocacy.

Air pollution is now a much discussed and feared issue. We are the most affected as half of the most polluted twenty cities in the world are in India. A recent study found that poor air quality causes 1.1 million premature deaths every year in India and huge health issues and costs for our people. The causes of air pollution are not well understood, so more research is required to identify the same so that this menace can be tackled. Better measurement and forecasting can allow for temporary timely actions even as we look for long-term solutions. Beijing forecasts a rise in pollution two weeks in advance so that the local government can take appropriate action. As citizens and media insist on transparency and data through measurement, we must make ourselves aware of the problem and then demand action from government and players who can do something about it.

We have seen how children have driven change over Diwali by saying no to crackers and in hygiene by WASH training in schools. Children are a great agent of change as they bring best practices learnt at school back to the home and community.

Industry pushed back on legislation on auto emissions. The courts had to intervene to address these issues and for India to move to Euro VI compliance by 2020. Industry too must embrace high standards in place around the world more readily, and we as citizens can drive this by where we spend our money. One of the most exciting outcomes of energy-efficiency ratings of air conditioners was that the price gap between the high and low ratings narrowed as consumers began to spend on the higher rated energy efficient consumer durables recognizing the benefit from lower electricity bills. Measurement and communication of the benefits have an important role to play in converting our behaviour as we look to do the right thing but need the information that helps us make the right decisions.

LEED (Leadership in Energy and Environmental Design) and GRIHA (Green Rating for Integrated Habitat Assessment) ratings of buildings have shown positive results as consumers show preference and encourage more builders to build to higher environmental standards as they realize the benefit of savings from energy and water efficiency. As citizens and consumers, we can impact pollution and climate change by the choices we make; in the vehicles we drive, the fuel we use, the way we deal with waste and burning, the materials and manner in which we construct, the buildings we live in.

Media needs to be cognizant of the consequences that unwarranted or negative reportage can make. Even more important is the outreach that media provides through advertisements and articles in helping behaviour change and providing the rallying call to arms. I would like to see many more media and communications channels step up to address this need and become avid champions as they play a critical role in the amplification of any social movement. Television, radio, content streaming on our mobile phones and Bollywood films—all help change behaviour. The success of *Toilet: Ek Prem Katha* is an example of a film grossing high earnings while carrying the right message.

Success can be best achieved through a community-led process with local leaders such as panchayats and community groups, supported by political and bureaucratic leaders. Change must be people-centric and designed with the end user in mind.

Our faith leaders are important influencers of behaviour change, helping us reflect on what we can and should do.

Strong citizen engagement, particularly working with youth, is important—since the future leaders and champions of social change are the young and India can leverage the great demographic advantage of its vibrant youth. Their strength in numbers and their energy and receptivity to new ideas can be harnessed to create strong on-ground movements.

This brings me to the important subject of jobs. We can create a large number of jobs by encouraging people to work in the provision of public services. The individual may not want a job in sanitation but may be happy to be an entrepreneur—running a business. There are excellent examples of for-profit models in sanitation where women run community toilets in Dharavi, or run and maintain drinking water ATM machines and solar-charging stations for solar lanterns. We need more social entrepreneurs and enterprises that provide public services and do this for a living. In order to encourage this, we need a thriving ecosystem of finance through venture capital and angel financing and mentors to encourage the technologies and ideas. We need our municipalities and government to enable the private sector to step in to fill the gaps. We need NGOs and communities and the government to come together to make workable service provider solutions as in the case of pay-to-use community toilets in Dharavi slums. We need to skill and educate our youth to seize the opportunity and deliver much-needed services—making these jobs and businesses aspirational.

To achieve this we need finance. The ISC is working with partners to deliver microfinance for sanitation to self-help groups to enable building of toilets by poor families. Banks have typically been more comfortable lending to MFIs rather than entering direct lending at the grassroots. The RBI recognizes lending for sanitation as priority sector lending. However, banks need aggregators who can deliver small loans to borrowers, which they get financed in turn by the banks. Ideally, if banks were to lend directly the cost of financing could come down for borrowers.

For large projects in the renewables space, or financing of STPs, or financing community toilets, banks need to get comfortable with the viability of these projects. Incentive structures and policies to make these projects a reality are needed. The regulatory and legal structures need to be foolproof as politicians are known to get tempted to declare these services as free. This perceived risk heightens the risk premium for these projects making financing difficult, kills the business model and entrepreneurs are reluctant to venture into these areas. We also need equity financing through venture capital and social entrepreneur funding.

Waste treatment is yet another contentious issue. Most urban citizens can afford to pay a small amount of money to treat the faecal sludge we generate—₹800 per annum and ₹80–100 per month for a middle-class household to contribute to the RWA or building society for sewage treatment is entirely feasible. But we need viable FSTP operators who will build and run the plant for us. Water for horticulture/flushing/construction can be a useful by-product. Let us encourage such solutions and those who will provide us these services.

Many state governments have building regulations requiring institutions and buildings to comply with high standards of sewage treatment. However, these plants are not built properly and become defunct within the year as builders take shortcuts in their desire to cut costs. As citizens, we need to fix these accountabilities and insist that these aspects are not only looked into but also acted upon to ensure prosecution for non-compliance.

Our research institutions can work to encourage solutions. Our IITs have incubators that have the ability to do this. The Defence Research and Development Organization (DRDO) has given India the biodigester toilets implemented by the railways and entrepreneurs. We need many more such institutions that develop solutions in India for India, even as we adopt and adapt that which is working elsewhere in the world. Solutions need to be shared more readily, and where successful, replicated rapidly across the country.

The book is replete with examples of private sector and corporate engagement whether through CSR budgets or as a business. The role of the private sector is not just the funding but in implementation of projects and the running of business that provide essential services. The role of SMEs and not just big businesses is critical. To do this, the regulatory environment needs to enable, indeed welcome, businesses and entrepreneurs in these spaces. As repeatedly stressed through the book, with many examples, we need to partner and collaborate.

Ultimately, it is for each of us to move the agenda of the country forward. We can start by doing the right thing or stopping someone from littering or polluting and ensuring their compliance to regulations and engaging with our local community leaders and politicians on the issue of providing adequate facilities to us. Only collectively through the creation of a people's movement, can we achieve an India that is fit for our children to live in—or else we are doomed to sink into a cesspool of poor health, disease and filth. I, for one, believe we are on the right path and at a very important inflexion point which will determine the India we want to build.

Acknowledgements

This book is based on articles that have appeared in leading publications—*The Indian Express, The Economic Times, Hindustan Times, Mint, Businessworld, Business Today, The Hill* and *Financial Times*—and also chapters published in *Opportunity Beckons, Economy of Jobs, Swachh Bharat: A Clean India* and *Women and Society: The Road to Change*.

I would like to acknowledge the contribution of many good friends for co-authoring articles with me, whose knowledge and experience I treasure: Paul Polman, CEO, Unilever; Sanjiv Mehta, CEO, HUL; Madhu Krishna at the Bill & Melinda Gates Foundation; Manas Rath, Senior Advisor, BORDA; Nick Robins, Co-Director, UNEP Inquiry into the Design of a Sustainable Financial System; Krishan Dhawan, CEO, Shakti Sustainable Energy Foundation; and my husband, Rashid Kidwai.

The team at FICCI—Anshuman Khanna, Pragati Srivastava, Sakshi Arora, Monika Dhole, Rita Roy Chaudhury and Swapna Patil—provided invaluable assistance with research in green finance and water related areas. I am also grateful to the members of the NCE team—Ferzina Banaji, Helen Mountford, Joel Jaeger, Rajat Kathuria and Gireesh Shrimali—for their enthusiastic support and research reports that informed my thinking reflected in some of the chapters. I gratefully acknowledge the contribution of the talented team at the ISC—Sidhartha Das, Medhavi Sharma, Zara Juneja, Shipra Saxena and Sudeshna Maiti.

I thank Gloria D'Souza for her efficient support as I wrote and rewrote chapters, and Kapish Mehra, Yamini Chowdhury and the team at Rupa Publications India who have done an admirable job of editing the book and seeing it to its conclusion.

And lastly, my gratitude to my family, as always is beyond measure—Kemaya, Rumaan and Rashid for their frank and critical inputs and support. Rashid, for his amazing capacity to brainstorm and attention to detail, Rumaan for his keen sense of design and command of the language and Kemaya for providing the lens of youth and her consulting experience together with her interest in these areas.

And for all the friends and supporters who encouraged me along the way—a big thank you!

The royalties from the sale of this book will go to the ISC as it continues its effective work in aiding India's journey to become ODF and helps human well-being and better health for all.

ISC is a credible platform for partnerships working towards a common goal of achieving sustainable sanitation for India, and has successfully brought together government, private sector, NGOs, donors, civil society and citizens to create synergies in sanitation. ISC works towards BUMT, thus addressing the entire sanitation value chain—bringing attention to construction of quality toilets and their use, their maintenance and treatment of the waste.

So thank you for buying the book—you have also contributed to a good cause!

Glossary of Important Terms

AKDN	Aga Khan Development Network
AWS	Alliance for Water Stewardship
AOL	Art of Living
ADB	Asian Development Bank
AIIB	Asian Infrastructure Investment Bank
AMRUT	Atal Mission for Rejuvenation and Urban Transformation
BCC	Behaviour Change Communication
BMGF	Bill & Melinda Gates Foundation
BRICS countries	Brazil, Russia, India, China, South Africa
BUMT	Build, Use, Maintain and Treat
BCS	Business and climate summit
CURE	Centre for Urban and Regional Excellence
CSR	Corporate Social Responsibility
GTZ	Deutsche Gesellschaft für Technische Zusammenarbeit
EY	Ernst & Young
EXIM	Export-Import Bank of India
FSM	Faecal Sludge Management
FSTPs	Faecal Sludge Treatment Plants

FBOs	Faith-based Organizations
FICCI	Federation of Indian Chambers of Commerce and Industry
FLO	FICCI Ladies Organisation
GCF	Green Climate Fund
GIWA	Global Interfaith WASH Alliance
HUL	Hindustan Unilever Limited
HSBC	Hongkong and Shanghai Banking Corporation
IL&FSE	Infrastructure Leasing & Financial Services-Education
ISC	India Sanitation Coalition
IWSN	India Water Stewardship Network
IREDA	Indian Renewable Energy Development Agency
IGCS	Indo-German Centre for Sustainability
IDRC	International Development Research Center
IEA	International Energy Agency
IWMI	International Water Management Institute
IRC-IWSC	International Reference Centre Community Water Supply and Sanitation—International Water and Sanitation Centre Japan International Cooperation Agency
L&T	Larsen & Toubro
MHM	Menstrual Health Management
MDGs	Millennium Development Goals
MDWS	Ministry of Drinking Water and Sanitation
MoUD	Ministry of Urban Development
NIIF	National Investment and Infrastructure Fund
NSDC	National Skill Development Corporation
NTPC	National Thermal Power Corporation

NCE	New Climate Economy
NDB	New Development Bank
NITI Aayog	National Institution for Transforming India
ODF	Open Defecation Free
RB	Reckitt Benckiser
SEBI	Securities and Exchange Board of India
SEWA	Self Employed Women's Association
SCGJ	Skill Council of Green Jobs
SPARC	Society for the Promotion of Area Resource Centers
SAIL	Steel Authority of India Limited
SDGs	Sustainable Development Goals
SDSN	Sustainable Development Solutions Network
SuSanA	Sustainable Sanitation Alliance
SBM	Swachh Bharat Mission
TERI	The Energy Resource Institute
USAID	U.S. Agency for International Development
UNEP	United Nations Environment Programme
UNEP Inquiry	United Nations Environment Programme Inquiry into the Design of a Sustainable Financial System
UNFCCC	United Nations Framework Convention on Climate Change
UNICEF	United Nations Children's Fund
UNCCC	United Nations Climate Change Conference
ULBs	Urban Local Bodies
WASH	Water, Sanitation and Hygiene
WBCSD	World Business Council for Sustainable Development
WEF	World Economic Forum
WWF	World Wildlife Fund

About the Author

Naina Lal Kidwai is Chairman of Max Financial Services and of the India Advisory Board of Advent Private Equity, and a Non-Executive Director of Cipla and Larsen & Toubro and the global board of Nestlé. She was the first woman to be elected as President of FICCI in 2013 and retired as Executive Director on the board of HSBC Asia Pacific and Chairman HSBC India in December 2015, having become the first woman CEO of a private bank in India in 2006.

Her environmental engagements include the UNEP International Advisory Council of the Inquiry for Sustainable Finance, Global Commission on Economy and Climate, Rockefeller Foundation Economic Council on Planetary Health, Shakti Sustainable Energy Foundation, Chair of FICCI's Sustainability, Energy and Water Council and Chair of the ISC. She is also committed to the empowerment of women and supports women's causes.

She is one of the five representatives from India on the BRICS Business Council and chaired the financial services working group; she is also a member of the INDO-ASEAN Business Council, the investment advisory committee of the Army Group Insurance, the Harvard Business School South Asia Advisory Board and the Governing Board of Lady Shri

Ram College. She has been a member of Government of India's Industry Task Force, the Prime Minister's Trade and Industry Council and the National Manufacturing Council.

The first Indian woman MBA from Harvard Business School, she appears in listings of international women in business and is the recipient of several awards including the Padma Shri from the Government of India.

She is an ardent lover of nature, arts and crafts, handlooms and Indian and Western classical music. Her family holidays are typically in the jungles with husband Rashid, her children, Kemaya and Rumaan, sister Nonita and nephew Zahaan.

Survive or Sink is her third publication after the bestselling *30 Women in Power and Contemporary Banking in India*.

Index

Achievable plans, 125–26
Acumen Fund, 52
Aditya Birla Group, 111, 117
Aga Khan Development
    Network (AKDN), 53–54
Agroforestry, 220
Air pollution, 109, 120–22,
    137–39, 141, 174–75, 181,
    184, 238
Al Gore Climate Reality Project,
    141
Alliance for Water Stewardship
    (AWS), 99, 101–02, 104
Anganwadi, 4, 19, 42, 220
Anti-national elements, xxxix, 155
Apple, 114
Arghyam, 23, 48
Art of Living (AOL), 69, 72
Asha Impact Trust, 52
Asian Development Bank (ADB),
    48, 85, 128, 165, 195
Asian Infrastructure Investment
    Bank (AIIB), 128, 195
Atal Mission for Rejuvenation
    and Urban Transformation
    (AMRUT), 62, 91, 176,
    178–79

Auto emissions, legislation on,
    238

Bagchi, Subroto, xxxv
Bal Swachhta Mission, 4
Bangladesh, 93–94
Behaviour change, xxiv–xxvi,
    xxviii, xl, 3, 5, 7–8, 10,
    22–23, 26, 30, 32–33, 37,
    40, 43–44, 48, 50, 53, 69,
    75, 233, 239–40
Behaviour Change
    Communication (BCC), 6,
    8, 24, 40, 45, 59, 232–33
Beti Bachao Beti Padhao, 210
Bhamashah Yojana, xl
Bharti Enterprises, 32
Bill & Melinda Gates Foundation
    (BMGF), 46, 48–49, 51
Biodigesters, xxix, 25
Biomass, 161, 171
Bloomberg World Index, 114
Bremen Overseas Research and
    Development Association
    (BORDA), 48
BRICS Business Council, xxxvii,
    191–92

BRICS countries (Brazil, Russia, India, China, South Africa), xxxvii–xxxviii, 139, 165, 188–95
BRICS Rating Agency, xxxvii, 191–95
BS IV fuels, 142
Build, Use, Maintain and Treat (BUMT), xxv, 6, 10, 22, 35, 37, 43, 61, 68, 74–78, 198, 202
Business and Climate Summit (BCS) 117, 134, 166
Business Opportunities for Women, 224–25

Cairn, 30
Capacity building, 104
Capital expenditure (CAPEX), 25–26, 67
Carbon footprint, 166, 181
Carbon sinks, 198, 200
Catchment areas, 15
CDP Climate Leadership Index, 114
Center for Policy Research (CPR), 55
Central Pollution Control Board (CPCB), 75–76, 92–93, 144
Central Rural Sanitation Programme (CRSP), 4, 57
Centre for Urban and Regional Excellence (CURE), 46
China, xxxvii–xxxviii, 93, 116, 121, 136–140, 167, 176, 186, 188–89, 192
  air pollution, 138
  ecological civilization, 137
  economic growth, 167
  GHG emitters, 137
  Green bonds, 189
  green financial system, 186
  investment in green industry, 186
  low-carbon pilot cities, 138
  odd–even traffic restrictions, 176
  renewable energy markets after, 162
  renewable energy, 140, 162
  solar and wind energy plants, 140
Citizen Science Leaders (CSLs), 97
Civil Society, xxx, 10, 96, 104, 132, 238
Clean Energy Fund, 180
Climate change, 109–10, 112–15, 118–19, 129, 131–34, 139, 164–66, 168, 185–88, 190, 196–97, 220, 233–34, 239
Climate Change Fund (CCF), 165
Climate commitments, 114
Climate Finance Lab, 164
Climate-friendly policies, 134
Climate-friendly solutions, 132–33
Climate Policy Initiative, 119, 125
Climate-sensitive sectors, 132
Climate Summit in Paris, 117
Closed-loop sanitation systems, 66
CLTS Foundation, 54–55
Coal-fired power plants, 111
Coca-Cola, 102, 115
Collaboration across stakeholders, xxx
Collaborative Drug Discovery,

Inc. (CDD), 48
Community Approaches to Total Sanitation (CATS), 47
Community-led approaches, 8
Community Led Total Sanitation (CLTS), 11, 11, 54–55
Companies Act, 2013, 21, 215
Compliance-driven approach, 21
Cooperative initiatives, 115
Corporate engagement, 25
Corporate Facilitation Desk, 66
Corporate Good Practices, 30–34
Corporate Governance, 214, 218
Corporate Helping Hands, 27
Corporate Partnerships, 225
Corporate sector participation, 37
Corporate Social Responsibility (CSR), xxvii, xxxi, 7, 15, 21–23, 26, 29–30, 33–38, 59, 156, 237, 242
Corporate volunteering, xxv, 42–43
Corporate water stewardship, 99
Council on Energy, Environment and Water (CEEW), 198
Credit decisions, xxxiii

Darwaza Band campaign, 4
Defecating in open, hazards of, xxiv–xxv
Defence Research and Development Organization (DRDO), 242
Dettol Banega Swachh India campaign, 32
Deutsche Gesellschaft Für Technische Zusammenarbeit (GTZ), xxvii, 47, 52, 62, 65
Development banks, 128
Development Linked indicators (DLI), 45
Development of Women and Children in Rural Areas (DWCRA), 222
Dharavi slums, xxviii, 240
Diarrhoea, xxiii, 9, 70, 93, 96
Diarrhoeal deaths, 13, 58
Disability-adjusted life years, 58
Domestic waste water, 19

E. coli, 83
Earthwatch, 97
Ease of Doing Business, 37
Ecological Civilization, 137–38
Ecological footprint, 133
Economic growth, 13, 48, 56, 60, 109, 114, 116, 118–20, 126, 137–38, 167, 208
Economy of Jobs, 201–02
Education, Women, 209–10
  Gross Enrolment Ratio, 209
  MDG of attaining universal education, 210
Electricity Act, 2003, 172
Employment generation, 8, 199
Empowered Working Group (EWG), 4
Energy Conservation Building Code, 180
Entrepreneurship, role of, xxxv
Entrepreneurship Skills, 202–04
Environmental indices, 116
Eradication of manual scavenging, 26
Ernst & Young (EY), 32, 48, 162
Euro VI compliance, 238
European Investment Bank (EIB), 162
Export-Import (EXIM) Bank, 169

Fabindia model, 225
Faecal Sludge, xxvi, 14, 17, 75, 79–81, 84–86, 202, 241
Faecal Sludge Management (FSM), xxvi, 48–49, 75, 79, 80–81, 85, 88–91
Faecal Sludge Treatment Plants (FSTPs), xxvi, 82–91, 203
Faith-based organizations (FBOs), xvii, xxv, 69–70, 73
Fast-moving Consumer Goods (FMCG), 22, 36
Federation of Indian Chambers of Commerce and Industry (FICCI), xxi, 53, 96, 101, 112, 134, 163–64, 166, 170, 181, 186, 212, 217–18
FICCI Ladies Organisation (FLO), 212, 217–18
FICCI Socio Economic Development Foundation (SEDF), 218
FICCI Water Mission, 96, 101
FICCI-India Sanitation Coalition Award, 53
Financial crisis, 185
Financial inclusion, xl, 52, 211–12
Financial Stability Board, 116, 187
FLO Women Empowerment (WE), 218
Fluoride problems, 94
Foreign direct investment (FDI), 114
Fortune 100 companies, 114
Fossil fuel subsidies, 138–40, 190
Funding, xxxiii–xxxiv

G20, xxxvii, 116, 138–39, 187
GAIL, 111
Gender Budget Statement, 235
Gender budgeting, 235
Gender equality, 60, 207, 217, 219, 231
Gender gap, 207–08, 215, 231
Gender in Policies, 234–36
Gender inequality, 207
Geographic Information Systems (GIS), 179
German Development Cooperation, 61–62, 64
Gir (Gujarat), xxxix
Glass ceiling, 219
Global Citizen India, 12
Global Commission on the Economy and Climate, 119, 125
Global Finance, 185
Global Gender Gap Report, 207
Global Interfaith WASH Alliance (GIWA), 69–71, 73
Global Warming, 185
Goa Declaration, 192
Godrej, 111, 117
Goldman Sachs, 208
Google, 114
Gramalaya Urban and Rural Development Initiatives and Network (GUARDIAN), 52
Green Agenda, 169–72
Green Bank, 165, 187
Green bonds, xxxiv, 112–13, 116, 129, 163–64, 169–70, 186, 189
Green Bonds Council, 164
Green Businesses, 198
Green Climate Fund (GCF), 165, 181

Green Conscience Keepers, 220–21
Green Economy, 159, 233–34, 236
Green Energy, 167–68
Green Finance, 127, 161, 164–66, 237
Green financial system, 116, 186
Green investments, 165, 170, 181
Green jobs, xxxiv–xxxv, 196–97, 200, 233
Green market, 168, 172
Green Rating for Integrated Habitat Assessment (GRIHA), 239
Green Revolution, 228
Greenhouse gas (GHG), 111–12, 117, 131, 186
Groundwater resources, 93
Growth model, xxxvi
Growth Potential of Green Businesses, 198–200

Habitat Conservation, xxxviii-xl, xxxviii
Handwashing campaign, 40
Havells, 30
Hepatitis E, 83
Hero Bicycles, 225
High-carbon economy, 109
High-carbon growth, 186
Higher Education for Women, 225–26
Hindustan Unilever Limited (HUL), xxviii, 30–31, 39–41, 43, 225
Honeysucker service providers, 90
Hongkong and Shanghai Banking Corporation (HSBC), xxi, xxxi, 31, 33–34, 97, 181, 220–21
Household Balance Sheet, 222
Hydro energy, 161
Hygiene Index programme, 32

ICRIER, 119
IFFCO, 27
Independent Verification Agent (IVA), 4
India Innovation Lab for Green Finance, 127
India Sanitation Coalition, xxii
India Water Stewardship Network (IWSN), 102
Indian Banks Association (IBA), 182
Indian Corporate Average Fuel Consumption Standard, 180
Indian Forest Service (IFS), 153
Indian Oil, 111
Indian Railway Finance Corporation, 27
Indian Railways, 40, 89, 111, 125–26, 145
Indian Renewable Energy Development Agency (IREDA), 165, 171
Indian Sanitation Coalition (ISC), xxv, xxvii, xxxiv, xl, 6, 10, 12, 20–21, 30–31, 35–38, 40, 42–43, 48–51, 62–63, 67, 70, 74, 77, 156, 198, 202–03, 218, 237, 240
Indo-German Centre for Sustainability (IGCS), 62–63
Indo-German partnerships, 67
Industrial waste water, 19, 93
Information, Education and

Communication (IEC), 24, 58, 233
Infosys, 111
Infrastructure Development Finance Company (IDFC), 51, 94, 235
Infrastructure Leasing & Financial Services-Education (IL&FSE), 30, 50
INSIGHTS, 51
Institutional investors, 127
Instruments for Green Finance, 164–65
Insurance Regulatory and Development Authority (IRDA), 168
Integrated Watershed Management (IWM), 95
Intended Nationally Determined Contribution (INDC), 196–97
International Development Research Centre (IDRC), 51
International Energy Agency (IEA), 115
International Labour Organization (ILO), 210
International Solar Alliance (ISA), xxxvii, 139
International Water Management Institute (IWMI), 52
IRC International Water and Sanitation Centre, 51
Islam, xxv
Islamic Relief India, 69–70
ITC, 27, 102–03, 117, 225
Iyer, Parameswaran, xxviii

Jain Irrigation, 102

Janalakshmi Financial Services Private Limited, 48
Japan International Cooperation Agency (JICA), 48
Jawaharlal Nehru National Urban Renewal Mission (JNNURM), 177, 182
Jhumming, 149
JSW Steel, 102

Kamal Kar, 54
Kaziranga, xxxix, 148–51
Knowledge-sharing, 136
Kohler, 30, 199, 202
KPMG, 170

Labour Force Participation Rates (LFPRs), 231
Lancet journal, 109
Land Value Capture Financing policy, 180
Larsen & Toubro (L&T), 27, 33
Leadership in Energy and Environmental Design (LEED), 239
LED lighting programme, 135
Low Carbon Technology Partnerships, 115
Low-carbon business models, 113
Low-carbon economy, 112–13, 116, 190, 230
Low-carbon growth, 114–16
Low-carbon solutions, 116
Low-carbon transition, 116
LPG programme, 135

Mahatma Gandhi National Rural Employment Guarantee Act (MGNREGA), 214, 226

Mahatma Gandhi, 39, 57, 130, 137, 214
Mahindra & Mahindra, 31, 102–03, 111, 117
Man–animal conflict, xxxix, 148, 153
Mann Deshi Udyogini, 226
Manual scavenging, 17
Mar del Plata Resolution (1977), 57
Mata Amritanandamayi Math, 27
Maternal and child health and safety, 9
Maternity Benefit (Amendment) Bill, 2016, 213
McKinsey, 126, 207
Menstrual Health Management (MHM), 9, 59
Microenterprise, 223–24
Microfinance, 223–24
Microfinance institutions (MFIs), 11, 223, 241
Millennium Development Goals (MDGs), 12, 14, 59, 230
Ministry of Drinking Water and Sanitation (MDWS), xxviii, xxxii, 3, 9, 11, 20–21, 26, 49, 57–58, 63–64, 75, 94, 233, 235
Ministry of Environment, Forest and Climate Change (MoEFCC), 144
Ministry of Finance (MoF), 179
Ministry of Labour and Employment, 222
Ministry of Urban Development (MoUD), xxviii, 3, 17, 26, 28, 63, 94, 179
Ministry of Women and Child Development (MoWCD), 235
Mobile-led rural BCC model, 40
Modi government, xxiv
Modi, Narendra, xxii–xxiii, 7, 74, 96, 110, 136, 168, 216
Monitoring methodologies, 95

National Ambient Air Quality Standards (NAAQS), 142
National Documentation Programme, 212
National Institute of Urban Affairs (NIUA), 46
National Investment and Infrastructure Fund (NIIF), 128
National Mission for Empowerment of Women, 235
National Plan for Women, 214
National Rural Health Mission (NRHM), 227
National Service Scheme (NSS), 208
National Skill Development Corporation (NSDC), xxxiv, xxxv, 199–200
National Thermal Power Corporation (NTPC), 34, 111, 144, 169
National Urban Sanitation Policy (NUSP), 58, 62
Nationally Determined Contributions (NDCs), 131
Natural Resources Defence Council (NRDC), 198
Nepal, 93, 231
Nestlé, 102
New Climate Economy (NCE),

xxii, xxv, 113, 118–19, 121, 125–26, 173, 187
New Development Bank (NDB), xxxvii, 128, 165, 189, 191, 193, 195
NFFSM Alliance, 49
NGOs and Development Agencies in Sanitation, xxvi

Nigrani Samitis, 5
Nirmal Bharat Abhiyan (NBA), 5, 57–58
Nirvana Fund, xxxv
NITI Aayog, xl
Non-fossil based electricity generation, 196
Non-green labour market, 236
Nuclear Power Corporation, 27

On-site sanitation (OSS), 17, 76, 79, 83, 85, 90
Open Defecation Free (ODF), xxiv, 3–7, 14, 32–33, 43, 45–47, 49, 53, 64, 68, 75, 78, 199, 203, 232–33, 237
Open defecation, elimination of, 26
OPEX (operating expenses), 26

Paris Agreement, xxxvii, 112–13, 124, 130–32, 137, 168, 188–90, 197
Pay-as-you-go solar home systems, 129
PepsiCo, 102
Perform, Achieve and Trade (PAT) scheme, 126
Pfizer, 115
Poaching, xxxix, 148, 153, 155–56, 158

Policy Actions, 122–24
Pollution, xxvii, xxxii, xxxiii, xxxvi, 14, 17, 60, 65, 75, 82, 84–85, 89, 94–95, 99, 104, 107, 109, 111, 120–21, 141, 143, 145–46, 175, 180, 196, 237–39
  emission standards for thermal power plants, 144
  key area, 145
  municipalities' treatment of waste and disposal, 146–47
  parking pricing policy, 145
  positive measures, 142–43
  remedial measures, 141
Poo2Loo campaign, 47
Pope Francis, xxv
Population Foundation of India, 59
Power outage (2012), 124
Pradhan Mantri Awas Yojana-Urban (PMAY-U), 177
Pradhan Mantri Bima Suraksha Yojana, 211
Pradhan Mantri Jan-Dhan Yojana, xl, 211
Pradhan Mantri Kaushal Vikas Yojana (PMKVY), 201
Premature deaths, 109, 111, 121, 137, 175, 238
Private investments to green industries, 170
Procter & Gamble (P&G), 115
Prohibition of Employment as Manual Scavengers and their Rehabilitation Act, 2013, 84
Project Tiger, xxxviii, 150, 158
Public–Private Partnership (PPP) models, 35, 66

Public Sector Undertakings (PSUs), 34, 43, 89, 111

Rainwater harvesting, 95, 102, 221
Ramakrishna Mission, 72–73
Reckitt Benckiser (RB), 30, 32
Remedial measures, xxxii, 141
Renewable energy, xxxiii, 109–10, 112, 114, 116, 125, 127, 131, 137, 140, 161–62, 169–70, 172, 189, 196–98, 200
Renewable Energy Certificate (REC), 172
Renewable energy investments, 162
Renewable Purchase Obligation (RPO), 172
Research Institute for Compassionate Economics (RICE), 55
Reserve Bank of India (RBI), xxxiv, 163, 168, 171, 241
Resource mismanagement, xxix, 94
Right to Education (RTE), 226
RO/water purifier plants, 25
RUDI, 225
Rural Sanitation, xxviii–xix, 4, 32, 43, 46, 57, 60, 72

Saaf & Safe for students, 50
SABMiller, 102–03
Safe drinking water, 31, 94
Safety for Women, 216–17
Samajhdar (Smart) campaign, 49
Sanitation and Hygiene, Advocacy and Communication Strategy (SHACS), 47
Sanitation budget, 91
Sanitation crisis, 13–14, 59
Sanitation divide, 18
Sanitation facilities for Women, 215–16
Sanitation plus, 66, 74
Sanitation-focused government programme, 59
Satya Bharti Abhiyan, 32
Sector Skill Councils (SSCs), 200
Securities and Exchange Board of India (SEBI), 163, 168, 215
Self Employed Women's Association (SEWA), 212, 224–26
Self-help groups (SHGs), xl, 211, 221, 240
Septage management, 18
Sewage treatment plant, 79, 82, 89
Sexual harassment policies, 216
Shakti Sustainable Energy Foundation, xxii
Shared Learning, 63
Shareef, Akmal, 70
Sharma, Rakesh, 141
Simhastha Kumbh Mela, 71
Singh, Birender, 10
Skill-based Training, 212
Skill Council of Green Jobs (SCGJ), xxxiv, 200
Skill development, 200–201
Skilling, xxxiv–xxxv, 200, 202
Sludge-emptying services, 79
Small and medium sized enterprises (SMEs), xxvii, xxxi, 35, 208, 242
Smart Cities, 8, 138, 176, 179, 181

Soak pits, 83
Social conflicts, 93
Society for the Promotion of Area Resource Centers (SPARC), xxviii, xxxv, 54, 203
Socio-economic displacements, 95
Solar power, 110, 124, 127, 140, 161–62, 169
Solid and liquid waste management (SLWM), 16, 19–20, 46
Solid waste, 3, 19, 26, 28, 33, 56, 58, 86, 176, 178, 198
Sony, 115
Stakeholders, xxx–xxxiii, 104–05
State Bank of India (SBI), 162
Steel Authority of India Limited (SAIL), 89
Steering Water Stewardship Goals, 101
Sulabh International, xxviii, 54, 218
Sulabh Shauchalaya, 54
Sustainable Development Goals (SDGs), 12, 14, 59–60, 118, 129–230
Sustainable Development Solutions Network (SDSN), 115
Sustainable financial ecosystem, 172
Sustainable Management of Water, 232–33
Sustainable Sanitation Alliance (SuSanA), xxvii, 11, 47, 62–63
Swachchata Kranti, xxv, 71–72
Swachh Aadat (Clean Habit), 30, 40

Swachh Bharat Abhiyan, xviii, 31, 216
Swachh Bharat Cess, xviii, 74
Swachh Bharat Fellows, xxxi, 49
Swachh Bharat Kosh, 21, 27, 33
Swachh Bharat Mission (SBM), xvii, xxii–xxiii, xxiv, xxvi, 3–7, 10, 12, 19–22, 25–30, 34–35, 38, 40–42, 45–46, 49–50, 54–56, 59–60, 62–67, 74–76, 78, 81–82, 91, 177–78, 198, 202, 236–37
Swachh Iconic Places, 34
Swachh India, 12
Swachh portal, xvxiii, 28
Swachh Survekshan survey, 181
Swachh Vidyalaya Swachh Bharat Abhiyan, 31
Swachhagrahi volunteers, 3
Swachhata Augmentation, 27
Swachhata Doot (Messenger of Cleanliness), xxv, 30–31, 39–43
Swachhata Saptahs (cleanliness drives), 24
Swachhata Status Report, 60

Tamil Nadu Water and Sanitation Pooled Fund, 183
Tata Group, 102–03, 111, 117
Tata Motors, 117
Tata Trusts, xxxi, 49
Tax-free bonds, 169, 179
Technical and Vocational Education and Training (TVET) Reform Project, 210
Telecom Regulatory Authority of India (TRAI), 41
Teresa, Mother, 43

The Energy Resource Institute (TERI), xxii
Toilet: Ek Prem Katha, 239
Toilet-building programme, 22
Total Sanitation Campaign (TSC), 4, 57
Tropical Forest Alliance 2020, 115

UK-India Infrastructure Fund, 181
Ultra Mega Power Projects (UMPPs), 167
UN climate negotiations, 134
Uncoordinated planning, 110
UNEP Inquiry, xxii, 111, 116, 163, 170
UNEP Inquiry India Council, xxii
United Nations Children's Fund (UNICEF), 16, 19, 46–47, 72
United Nations Climate Change Conference (UNCCC), 161
United Nations Environment Program (UNEP), xxii, 111, 116, 129, 163, 170, 181, 186, 188
United Nation Framework Convention on Climate Change (UNFCCC), 132, 135, 165
Universal access to toilets, 17
Urban Local Bodies (ULBs), xxviii, 3, 26, 28, 177–81, 183–84
Urban Management Center (UMC), 46
Urban Sanitation, xxvii–xviii, 15–16, 62
Urbanization, 137
Urban–rural divide, 24–25
U.S. Agency for International Development (USAID), 32, 46, 59
Usha Sewing Machines, 225

Value-driven approach, 21
Vector-borne maladies, 17
Vermicomposting, 220
Village Water and Sanitation Committees (VWSC), 236
Villgro, 52
Vishakha Guidelines, 216, 218

WASHCost, 51
Waste management, xxxiv, 16–20, 35, 52, 54, 64, 67, 186, 198, 200, 202
Waste-to-energy model, xxxv, 65
Waste treatment, xvi–xvii, xxxv
Water and Sanitation Decade, 234
Water for People, 48
Water Resources Group, 92
Water stewardship, xxi, xxx, 99–104, 238
Water, Sanitation and Hygiene (WASH), xxix, 20–24, 30, 33–34, 38, 43, 46–47, 51, 71, 234, 238
Water.org, xxxiv, 11, 50
WaterAid, 50, 97
Waterbodies pollution, 89
Waterborne diseases, 17, 58, 60, 81, 94, 96
Water-stressed economy, 92
Wildlife, xxxi, xxxviii–xl, 148, 150, 156–57
Wildlife Coalition, 156

Wildlife Habitats
  Anamalai Hills, 154
  inbreeding, 149
  internal conflict within the forest service, 153
  lions in Gir, 149
  man–animal conflict, 148
  organizational, administrative and monitoring costs, 149
  park management, 157
  park visits and education on conservation, 157
  poaching and animal trade, 155
  Ranthambore, 152
  rhino conservation in Kaziranga, 148–49
  tourism, 157
  wildlife coalition, 156–58
Wildlife Protection Act (1972), xxxviii, 150
Wind energy, xxxiv, xxxvii, 140, 161, 198
Wipro, 111

Women Empowerment, 212, 219–28
Women in Corporate Governance, 214–15
Women's participation in workforce, 208
Women's Reservation Bill, 214
Women-specific financial services, 211
Workplace Infrastructure for Women, 213–14
World Bank, xxiv, 4, 45, 60, 93, 111, 128, 162, 195, 232
World Business Council for Sustainable Development (WBCSD), 101, 115
World Economic Forum, xxii, 207, 231
World Resources Institute (WRI), xxix, 93, 119
World Wildlife Fund (WWF), xxxi, 97, 101

Zero Liquid Discharge (ZLD), 95